T0220868

QUICK CARE

AND DYING WISHES

QUICK CATTLE AND DYING WISHES

PEOPLE AND THEIR ANIMALS IN EARLY MODERN ENGLAND

ERICA FUDGE

CORNELL UNIVERSITY PRESS

Ithaca and London

First published 2018 by Cornell University Press

Library of Congress Cataloging-in-Publication Data

Names: Fudge, Erica, author.
Title: Quick cattle and dying wishes : people and their
    animals in early modern England / Erica Fudge.
Description: Ithaca : Cornell University Press, 2018. |
    Includes bibliographical references and index.
Identifiers: LCCN 2017048071 (print) | LCCN 2017051601
    (ebook) | ISBN 9781501715099 (epub/mobi) | ISBN
    9781501715105 (pdf) | ISBN 9781501715075 | ISBN
    9781501715075 (cloth) | ISBN 9781501715082 (pbk.)

Subjects: LCSH: Human-animal relationships—England—
    History—17th century. | Domestic animals—England—
    History—17th century. | Wills—England—History—
    17th century. | England—Social life and customs—
    17th century.
Classification: LCC QL85 (ebook) | LCC QL85 .F83
    2018 (print) | DDC 591.942—dc23
LC record available at https://lccn.loc.gov/2017048071

❧ Contents

✒ PREFACE

Looking for Animals in Early Modern England: A Note on the Evidence

Agricultural animals were a constant presence in the lives of a large number of people in early modern England, and it is the relationships formed between those people and their animals are the focus of this book. Despite their being commonplace, however, there is a difficulty: little evidence exists about the nature and quality of those relationships or the lives of either the animals or those who worked with them. Not only were many of the people who worked most closely with animals illiterate, the utterly prosaic nature of the face-to-face encounters between humans and livestock also meant they were rarely written down: what happened without thinking perhaps passed without recording. Husbandry manuals and early animal health care texts offer some insight, but even these are limited in the perspectives they offer. By their nature, husbandry manuals and animal health care texts are generic: they do not deal with actual face-to-face encounters between individual humans and specific animals that made up so constant a part of early modern life for so many, and it is the ordinary and the particular that are the focus here. Perhaps paradoxically, it seems that when what is ordinary is interrupted, the everyday comes most fully into view, and for this reason, I have used wills, documents that mark the ultimate interruption of the norm—death—as key sources in what follows.

At the heart of this book is a dataset of 4,444 wills from the Essex Record Office (ERO). This dataset (which I refer to as the large ERO dataset) covers the period 1 January 1620 to 31 December 1634 and includes wills from residents of Essex.[1] These wills come from five different classifications in the ERO archives: 2,275 wills classified as ABW are from the Commissary

---

1. I have changed the beginning of a year throughout from the old start date of 25 March to the modern date of 1 January for the sake of simplicity. Thus, a document originally dated, say, 16 February 1627 is represented here as 16 February 1628.

of the Bishop of London; 929 documents classified as ACW are from the Archdeaconry of Colchester; 855 documents classified as AEW are from the Archdeaconry of Essex; 351 wills classified as AMW are from the Archdeaconry of Middlesex (Essex and Hertfordshire Jurisdiction); 25 documents classified as APbW are from the Archbishop of Canterbury: Peculiar of Deanery of Bocking; and the smallest group of nine wills are classified APgW, from the Bishop of London: Peculiar of Good Easter. I have excluded all wills in the ERO in this period from testators from outside Essex.[2] Other wills which have been excluded from the dataset are damaged or incomplete.

This large ERO dataset offers data on a range of general issues—on place of residence, gender and occupation/status of testator, year (the probate date is used for this), whether the document is a formal written will or a nuncupative one (i.e., a transcription of a deathbed statement attested to by witnesses),[3] and whether it is signed or marked by the testator. In addition, the dataset records the nature of the bequests in each will, which I have classed under the broad headings of property (i.e., real estate), money, goods, and animals.

I have constructed a further dataset from the large ERO dataset. This sample of eighty-nine wills (which I refer to as the sample dataset) is made up of every fiftieth will from the large ERO dataset, which was ordered alphabetically by classification, and then numerically. More detail is offered by this dataset on issues such as the length of the documents, the kind and detail of the collecting phrase at the will's end (discussed in chapter 1), who the executor was, and who the legatees. Inevitably, the book pays closest attention to wills that include specific animal bequests, by which I mean not only bequests of animals that are named or described in individual detail— "my red cow," for example—but also in general terms such as "all my sheep" or "my cows."

In chapter 5, in addition to this archival data from Essex, I also introduce a comparative dataset of 2,013 wills from London and Middlesex from the Diocese of London, held in the London Metropolitan Archive (LMA) for

---

2. In the period of the dataset, the majority of excluded AMW wills were from Hertfordshire. Others were from Cambridgeshire, Hereford, Huntingdonshire, London, and Middlesex.

3. Throughout, when discussing individuals "writing" their wills, I am using the term to describe what was often the process of dictating the document's content to a scribe. Nuncupative wills record a deathbed statement as heard by the witnesses (audience) and are described here as "spoken" by the testator to distinguish them.

the same period. These wills (which I refer to as the LMA dataset) come from the classifications MS 9172/31 (1620) to MS 9172/42 (1634) (no wills from 1623 have survived in the archives) and includes 1,130 from testators who lived in the City of London and 883 from those in surrounding parishes in Middlesex. I excluded damaged and incomplete wills from this dataset as well as wills from outside London and Middlesex and wills by testators whose place of residence is not known. The wills in the LMA dataset offer basic data on the same range of general issues as those contained in the large ERO dataset.

I read most of the ERO wills from their originals, but due to constraints of time and distance, I read scans of some wills on the ERO's website, http://seax.essexcc.gov.uk (SEAX). I read all of the LMA wills from the originals, then used online scans on www.ancestry.co.uk (Ancestry) for final checking during the writing of the book (due, again, to issues of distance and access).

In addition, I have constructed a very limited dataset of a small sample of wills from the Prerogative Court of Canterbury (PCC), which had jurisdiction over wills by testators who had property in more than one location. This PCC dataset, which I mention briefly in chapter 5, is made up of 161 wills by testators who lived in Essex, London, and Middlesex that were made probate in 1627, the middle year of the larger datasets. I read scans of registered copies of these wills (that is, the official replicas made at the time) on Ancestry. The originals are held in the National Archive (NA), and I have used NA reference numbers to identify the wills I specifically refer to.

As well as using wills I have also looked at other manuscript holdings in the ERO, LMA and NA, including accounts and commonplace books. Records such as these afford another means by which some of the priorities of early modern people might be made available to us. In addition, I have used the parish registers that record baptisms, marriages, and burials in the period to track down, where possible, more details of the lives of testators and their legatees than are evident in wills. I read scans of these documents on SEAX and Ancestry. Where a detail is hard to decipher on an ERO scan, I have been able to look at the original. The scan quality is excellent, however, and I needed to do this very few times.

Although the wills are limited in what they endeavor to do (their intentions are very clear and very narrow), they can offer glimpses of the interactions between humans and animals that are otherwise invisible to history. But, at the same time, wills, as those who have used them before to study the past have recognized, do not offer a complete picture of early modern English society.[4] Made mainly by those in the upper and middle ranks, wills reveal very little evidence of the lives of those in the lowest stations in life.[5] In the large ERO dataset, for example, there are 96 gentlemen's, 1,121 yeomen's, and 737 husbandmen's but only 60 laborers' wills despite the fact that, according to Alan Everitt, such laborers made up "about one quarter or one third of the entire population of the countryside."[6] What this means is that those whose lives may have been most intimately tied up with the fewest animals might remain invisible: Thomas Dekker noted in 1615, for example, that the impact of the weather fell hardest on the "poor cottager": he "that hath but a cow to live upon must feed hungry meals (God knows) when the beast herself hath but a bare commons."[7] And, likewise, William Poole noted in 1650, "for the Poorer sort of people . . . having but one Beast, the losse thereof to them, is more then the losse of many to a Rich man."[8] In addition, because of their restricted position in relation to the ownership of property, women are underrepresented: in the large ERO dataset there are 3,559 wills by men and only 885 by women. As I will show in chapter 3, among the human household the housewife was likely to have the closest relationship with the dairy cows, but this closeness might remain invisible in wills as those herds were so often figured as the property of the husband. Also remaining invisible, of course, are those individuals who bequeathed property informally—that is, without writing a will. And there are a lot of such people—Ralph Houlbrooke notes one positive assessment that in the 1620s, "at least" 19 percent of the population left a will.[9] By implication, perhaps

---

4. See, for example, Richard T. Vann, "Wills and the Family in an English Town: Banbury, 1550–1800," *Journal of Family History* 4, no. 4 (1979): 346–67; W. Coster, *Kinship and Inheritance in Early Modern England: Three Yorkshire Parishes*, Borthwick Papers no. 83 (York: St. Anthony's Press, 1993): 1–2; and Ralph Houlbrooke, *Death, Religion, and the Family in England, 1480–1750* (Oxford: Oxford University Press, 1998), pp. 84–85.

5. There are exceptions: in wills from late Elizabethan Willingham, Cambridgeshire, Margaret Spufford found that "over three-quarters of the testators made a will, not because they were rich or poor, but because they had to provide for children who were not yet independent." Spufford, "Peasant Inheritance Customs and Land Distribution in Cambridgeshire from the Sixteenth to the Eighteenth Centuries," in *Family and Inheritance: Rural Society in Western Europe, 1200–1800*, edited by Jack Goody, Joan Thirsk, and E. P. Thompson (Cambridge: Cambridge University Press, 1976), p. 171.

6. Alan Everitt, "Farm Labourers," in *The Agrarian History of England and Wales*, vol. 4, *1500–1640*, edited by Joan Thirsk (Cambridge: Cambridge University Press, 1967), p. 398.

7. Thomas Dekker, *The cold year. 1614. A deepe snow: in which men and cattell haue perished* (London: W. W., 1615), B3v.

8. William Poole, *The Countrey Farrier* (London: Tho. Forcet, 1650), A3.

9. Houlbrooke, *Death, Religion, and the Family in England*, 84.

80 percent of the population did not.[10] Despite this, I hope that the evidence in the wills that we do have means that my chosen method of gaining insight into the world of the fields and the yards of early modern England is not wholly fruitless.

The wills and related primary materials provide the core data for this study, but in addition to them I am using some material that is rather less common in historical study: work from the fields of animal studies and animal welfare science (AWS). While the introduction of modern theoretical materials into the study of early modern cultures is hardly new, using work that is emerging from the study of the lives of modern livestock animals brings risks beyond those encountered when applying, say, poststructuralist theory to seventeenth-century culture. Is a cow today anything like a cow from the first half of the seventeenth century? How do the physical changes that have taken place—with in-and-in breeding, with improvements in the understanding of nutrition, and so on—impact the animal's behavior? I argue here that modern analyses of, for example, the collaborative nature of humans' work with livestock animals in agricultural production can inform the evidence that the wills and other early modern documents and texts provide; that contemporary understandings can enhance our comprehension of past human-animal interactions. There is, of course, a risk of being anachronistic, of applying ideas that find no echo in the earlier period, but I propose that the use of these modern materials that might offer further insight into what were vital relationships in the seventeenth century is worth the carefully managed risk. The early modern evidence is primary, but the findings of AWS and work from animal studies help me to ask new questions of early modern evidence. In addition, in the afterword, I suggest how we might read some findings from AWS as being less anachronistic than might initially seem to be the case.

This book has been written with two key audiences in mind. Most obviously, and primarily, the book is aimed at scholars of early modern culture and society. I hope that it will add another layer to our understanding of that world. But the book is also aimed at the growing group of academics and non-academics who are working in the field of animal studies. This interdisciplinary field, which has emerged in the last twenty years, engages with the various ways humans and animals have interacted. Ideas coming out of animal studies

---

10. It is worth noting that according to a 2010 survey, "almost two-thirds" of adults in the UK aged 35–54 and one-third of those over 55 do not have a will. "30m UK Adults Have Not Made a Will," *The Guardian*, 22 October 2010, https://www.theguardian.com/money/2010/oct/23/making-will-dying-intestate accessed 1 June 2017.

work most successfully, I think, when they engage with materials from a range of disciplines because animals, it has been recognized, are not confined to only one aspect of human culture. They exist in reality and in representation, and in the complex moments in between those two. Indeed, the anthropologist Garry Marvin has argued that "from the perspective of the humanities, the real animal *is* the cultural animal. Humans do not become more interesting or reveal themselves more truly when their cultural clothing is removed. . . . I think that the same holds true for nonhuman creatures."[11] Relations between people and their livestock in early modern England reflect this.

Because I wrote this book for both early modernists and animal studies scholars, on occasion, what might be obvious to an early modernist is given a gloss for non-specialist readers; familiar terms and ideas in animal studies are outlined in more detail than those working in that field require. I apologize for this, but hope that the rationale for such moments will excuse them. The dual readership can also, I suggest, be beneficial. One possible outcome of aiming at both is that early modernists will be introduced to some fascinating work in animal studies and that animal studies scholars will be offered insights into historical debates about human-animal relations. It is worth stating that history—like all disciplines—is, of course, interpretative, and I hope that the questions I am asking of my early modern materials might also inform what questions might be asked of any other materials pertaining to human-animal relations, historical or contemporary. What follows is not fact: the key arguments emerge from acts of interpretation of particular moments in particular documents and from the choreographing of ideas into what is, I hope, a coherent argument.

That people wrote wills with the intention of bequeathing their property to particular individuals after their death cannot be doubted, and the documents are frequently clear to the point of tedium for this reason. But why people bequeathed some things and not others, how they bequeathed things, and why they bequeathed what they did to whom is less obvious and is worth speculating about. Also significant is that on occasion wills, which are generally such conventional documents, contain moments of individuality. These can be the ways in which animals are described, or in the glimpses of particular domestic arrangements that we get, and it is these moments that form the core of this book. Thus, despite the sizeable datasets that underpin what follows, this book

---

11. Garry Marvin, "Wolves in Sheep's (and Others') Clothing," in *Beastly Natures: Animals, Humans and the Study of History*, edited by Dorothee Brantz (Charlottesville and London: University of Virginia Press, 2010), p. 75.

is not a piece of quantitative history. When it is used, the quantitative evidence is offered to illustrate a point that is then taken further with analysis of qualitative findings, but I hope that the quantitative details where they are used will be of interest. Scholars of early modern culture will note that much of the general data that emerges from the datasets (on literacy rates, for example) reinforces what other historians have shown using other sources and that this work will offer further evidence for established findings in the study of the period. But what the data used here will do that is new is afford an insight into the detail of human lives lived alongside animals. No one before has taken such a sizeable set of documents and used them to track the intertwined lives of people and livestock in the early modern period, and I would argue that this is a vital step for us to take if we are to write histories of early modern men, women, and children.[12] Many of those early modern men, women, and children spent many hours of every day, day-in, day-out, working with, living with, and probably worrying about their animals. If we are to have any understanding of those people and their lives, we need to try and contemplate what they contemplated, be concerned about what concerned them.

In addition, as will be obvious to anyone who has ever shared their life with an animal—be it a cow or a cat, a horse or a hamster—animals are not simply objects; they are sentient beings with the capacity to respond and resist, who can offer and receive affection. In tracking the lives that humans and animals shared in the early modern period, I hope also to recognize the active roles animals played.[13] The livestock animals of the first half of the seventeenth century, as before and as after, were not only weighed, appraised, counted; they were cared for, stroked, killed, spoken to, and some were even named. They were treated well and treated badly; they were individuals who had to be negotiated with, collectives who had to be encountered as such. The ways people lived with their animals will tell us much about those animals and about those people and the social and emotional worlds they shared.

---

12. Alan Mikhail makes some use of wills in *The Animal in Ottoman Egypt* (Oxford: Oxford University Press, 2014).

13. The book pays less attention to horses than other livestock animals for the simple reason that historical scholarship has already appeared on this species. See, for example, Joan Thirsk, *Horses in Early Modern England: For Service, for Pleasure, for Power* (Reading: University of Reading, 1978); Peter Edwards, *The Horse Trade of Tudor and Stuart England* (Cambridge: Cambridge University Press, 1988); Peter Edwards, *Horse and Man in Early Modern England* (London: Continuum, 2007); Sandra Swart, *Riding High: Horses, Humans and History in South Africa* (Johannesburg: Witts University Press, 2010); and essays in Karen Raber and Treva J. Tucker, eds., *The Culture of the Horse: Status, Discipline, and Identity in the Early Modern World* (Basingstoke: Palgrave, 2005) and in Pia F. Cuneo, ed., *Animals and Early Modern Identity* (Farnham: Ashgate, 2014).

Without having some understanding of this aspect of the past, especially when that past was so dominated by small-scale agricultural production, it is difficult to see how we can view our history of the period as complete.

The book is called *Quick Cattle and Dying Wishes* because it takes the link between the animals and the final testaments written in the period seriously. "Quick cattle," however, is not just a historically correct term for the animals that appear in wills: it reveals an issue that is central here. As chapter 1 will show, the term "cattle" was used in this period to mean all animals, occasionally cows in particular, and frequently just stuff (which might also be termed "chattels"). "Quick cattle" thus also embodies a question of perception: what the animals were—sentient beings, co-workers, objects—to the people who owned them. In addition, the book thinks about the nature of writing a will; it doesn't just use wills as blank sources that come from and go nowhere. Instead it takes seriously the impact of the documents and their intentions on understanding the animals they include. Crucially, and simply, the book argues that quick cattle were part of people's dying wishes and that that itself is significant and is something that needs to be considered.

Even as Laurie Shannon has shown that "animal" was not a common term in the early seventeenth century,[14] I have used it throughout and have not used "nonhuman animal" to signal a distinction from the "human animal" whose will might be read as evidence. This is mainly because "animal" has a non-abstract familiarity to us that "nonhuman animal" does not and because much of the time when I am writing about animals in this book I am also writing about non-abstract familiarity.

In transcribing manuscript documents, I have struck through parts of the will that are ~~crossed out~~ in the original, and where what is crossed-out is illegible that is represented as ~~xxxxx~~. When something has been added later above a line, it is presented ^thus^; added marginal comments are presented <thus>. Spelling and punctuation are as in the original documents, and on occasion this reveals the pronunciation of the speaker: "Roofe Barbor," a legatee in one will, for example, is likely to be Ruth Barber, but I have kept her as "Roofe," as the scribe represented her 400 years ago. I have modernized place names in wills for the sake of clarity, and I have used the first representation of the name of an individual in a will for subsequent spelling of their name even if it changes in the remainder of the will or is different in another document, such as a parish register.

---

14. Laurie Shannon, "The Eight Animals in Shakespeare; or, Before the Human,"*PMLA*, 124, no. 2 (2009): 472–79.

❦ Acknowledgments

This book began in 2008 as a study of perceptions of animal communication in early modern England, and changed focus when, in 2009, I was invited to give a talk at the Welfare Quality® conference "Knowing Animals" in Florence. As I was considering how to address an audience of veterinarians, animal welfare scientists, and agriculturalists in Italy, I was also teaching John Milton's *Comus* to students in London, and in the collision of those two things I came to wonder about Milton's use of the term "herd" for a group of creatures with no sense of self or home and whether it was a conception that would have been shared by the men and women who worked alongside animals at the time he was writing. The book is the outcome of that wondering.

I have many, many people to acknowledge for sharing their expertise over the lengthy period I have been contemplating this book. In particular, I would like to thank Mara Miele for that invitation to go to Florence and for lots of discussions since, Françoise Wemelsfelder for being willing to talk QBA with a historian, Sandra Swart for the historical exchanges, the reader's report and the tea, and Nigel Rothfels for the reports and suggestions. Conversations with Vinciane Despret have made me think better—and, I hope, more politely. Sarah Cockram and Jonathan Hope read chapters and offered invaluable comments, Richard Thomas took me on an archaeological dig and explained bones, Patricia Brewerton gave my paleography skills a polish, and Erik Fudge helped with the Latin.

Lots of other folk gave references, time, ideas, advice, friendship. Thanks go to Eleanor Bell, Donald Broom, Henry Buller, Carolyn Burdett, Jonathan Burt, Ali Cathcart, Douglas Clark, Susan Curry, Diana Donald, Holly Dugan, Sarah Edwards, Nigel Fabb, Sarah Franklin, Heather Froehlich, Andrew Gardiner, Kay Gilbert, David Goldie, Laura Gowing, Faye Hammill, Alan Hogarth, Martin Irwin, Elspeth Jajdelska, Matthew Johnson, Hilda Kean, John Law, Rebecca Marsland, Garry Marvin, Jennie McDonnell, Jim Mills, Lawrence Normand, Clare Palmer, Laura Paterson, Susan Pearson, Laura Piacentini, Mahesh Rangarajan, Ali Ryland, Stella Sandford, Steven Veerapen, John Webster, Rhoda Wilkie, and Sue Wiseman.

I am very fortunate to hold a meeting of the British Animal Studies Network at the University of Strathclyde each year and to go to another at some other exotic location in the UK. I thank all the participants for their animal thinking. Likewise, I am grateful to my comrades in the Glasgow Animal Studies Reading Group—Sarah Cockram (again), Rebecca Jones, Louise Logan, Karen Lury, Heather Lynch, Isabelle Pollentzke, Elsa Richardson—who have prodded my brain with their ideas.

Bits of this book have been given as papers at various locations over the past few years. I would like to thank the following for inviting me: Greg Bankoff at the University of Hull; Kristian Bjørkdahl for a memorable event at the Norwegian Festival of Literature in Lillehammer; Melissa Boyde at the University of Wollongong; Emily Brady at the University of Edinburgh; Ron Broglio at Arizona State University; Louise Hill Curth at the University of Winchester; Andy Gordon at the University of Aberdeen; David Harradine and colleagues in Fevered Sleep; Stephen Houston at Brown University; Thomas Laqueur and Alan Mikhail at the University of California, Berkeley; Susan McHugh at the University of New England in Maine; Bob McKay and John Miller at the University of Sheffield; Nicole Mennell and Jennifer Reid at the London Renaissance Seminar; Vivek Menon and colleagues at the Wildlife Trust of India; Brett Mizelle at California State University, Long Beach; John Mullarkey for an event at the Natural History Museum; Fiona Probyn-Rapsey and Dinesh Wadiwel at the University of Sydney; Peter Sahlins for an event at the Maison Européenne des Sciences de l'Homme et de la Société; Laurie Shannon at Northwestern University; David Gary Shaw at Wesleyan University; Kim Stallwood for the talk at Vegfest; Tom Tyler at the University of Leeds; Charles Watkins at the University of Nottingham; and Yvette Watt at the University of Tasmania. Thanks too to the people who asked questions and made suggestions at those events.

In addition, I am grateful to Joseph Campana and Cary Wolfe for hosting me at Rice University for a semester in the Humanities Research Center there, which gave me time to contemplate the emblematic worldview; and to Barbara Creed and colleagues at the University of Melbourne, where the writing actually started, for hosting me on the Macgeorge Fellowship in July 2015.

Despite the contributions of these people, there will be errors here and they are all my very own.

The archivists and staff at the Essex Record Office in Chelmsford were fantastically helpful, willing to celebrate with me when I found a cow with a name and to commiserate on the days when all I got was a solitary lamb. Since my move to Glasgow from east London, Katharine Schofield at the

ERO has been brilliant at answering queries, finding lost pages, and translating strange terminology via email. I'm very thankful to her. Staff at the London Metropolitan Archive and National Archives were also generous with time and advice.

I am grateful to the Wellcome Trust (grant WT101997MA) for funding that allowed me to finish the LMA dataset; to the School of Humanities at the University of Strathclyde for two periods of research leave; and to the AHRC for the Leadership Fellowship (AH/M008436/1) in 2015–2016 that allowed me the time to write the final draft of the whole book. Without it I don't think this book would ever have seen the light of day.

Mahinder Kingra at Cornell University Press has been supportive throughout and made comments that really helped the book to take the shape it has. I am grateful for his insight.

Finally, I'd like to thank Mum, Dad, Tessa, Tim, Julie, Osian, Oran and Macsen for appearing to be interested. This book is in memory of my uncle, Arthur Hunter, who, as he pointed out to me once, knew more about cows than I ever will.

❦ QUICK CATTLE
AND DYING WISHES

Introduction
Goldelocks and the Three Bequests

There is one cow who will not get into this book. Her name is Goldelocks and she appears in the will of Robert Jacobb, a yeoman—that is, a "substantial husbandman farming an acreage well in excess of that needed to support his family."[1] Her presence in this document and exclusion from this book forms the focus of this introduction, which will offer a reading of Jacobb's will in the context of trends traced in others in the large ERO dataset. The hope is that an analysis of Jacobb's final distribution of his possessions will reveal how wills work in this period: what they are likely to contain, what to omit. The analysis also allows for an explanation of how these documents are used in this book—what limits have been set and why.

Jacobb's will was written by a scribe on 6 November 1617, at which time Jacobb was, as the document notes, "of good and pfect memory my heavenly God therefore be praised."[2] Such a state was required of any testator. In *A Briefe Treatise of Testaments and Last Willes*—which became a standard work in the late Elizabethan period and after—Henry Swinburne wrote that "such personnes as haue not the vse of reason or vnderstanding, as madde folks, or idiots, are iustlye excluded from making of testamentes, . . . for their deuises

---

1. Keith Wrightson and David Levine, *Poverty and Piety in an English Village: Terling, 1525–1700* (Oxford: Clarendon, 1995), p. 6.
2. ERO, D/AMW 3/172 (1617), made probate 47 days after writing.

being full of folly, theyr deedes must needs be voyd of discreation and their wittes being sencelesse."[3] In addition to his possession of a good and perfect memory, we know quite a lot else about Jacobb from his will. He was a Protestant,[4] as is evident in the will's lengthy preamble to the bequests:

> I comitt and comend my soule into the handes of Almighty God my creator and maker, and renouncing all other hopes of salvacon doe only hope and confidently trust to be saued through the merrits and passion of Jesus Christ his only sonne and my only redemr, my body I will to be buried at the discreacon of my executors hereafter named.[5]

Once we get beyond the religious preamble, a version of which is included in virtually every will, we reach the bequests. As such, Jacobb's document is typical: after the statement of the testator's name, occupation (or rank if the testator was male, marital status if the testator was female); the name of the parish; the date the will was written; and confirmation of the testator's good memory, the religious preamble, with its focus on the eternal, marked the end of these conventional preliminaries. Following these, the will then turned to the temporal realm—to the world of material things. Here burial was mentioned (often by simply stating, as Jacobb did, that the testator wished their body to be buried at the discretion of their executor), and then the bequests would start. And these also tell us much about the testator.

We know from his will, for example, that Jacobb owed his brother Humphrey Spencer the large sum of £255, with which he had bought land "for

---

3. Henry Swinburne, *A Briefe Treatise of Testaments and Last Willes* (London: Iohn Windet, 1590), p. 8r. See also Sheila Doyle, "Swinburne, Henry (c.1551–1624)," *Oxford Dictionary of National Biography* (Oxford: Oxford University Press, 2004).

4. The religious feeling could be the scribe's rather than the testator's, of course. On this point, see Margaret Spufford, "Religious Preambles and the Scribes of Villagers' Wills in Cambridgeshire, 1570–1700," in *When Death Do Us Part: Understanding and Interpreting the Probate Records of Early Modern England*, edited by Tom Arkell, Nesta Evans, and Nigel Goose (Oxford: Leopard's Head Press, 2000), pp. 144–57. In a selection of twenty wills from Standon dating 1616–1619 in the Essex Record Office, it is difficult to link Robert Jacobb's will with a named scribe, but of these twenty wills, one Thomas Jacob (perhaps related to the testator Robert) wrote and witnessed five (signing himself as "scr"). Each of his five wills included a similar preamble: "I comend my soule vnto my Saviour & redeemer Christ Jesus by the merits of whose death & passion I trust to be for ever in his glorious kingdome, and I will my body to the earth to be buried in Christian buriall" (ERO, D/AMW 3/157 [1618]—made probate twenty-three days after writing). The scribe Jacob's other wills from Standon in this period are ERO, D/AMW 1/143 (1619), made probate sixty-nine days after writing; ERO, D/AMW 1/152 (1619), made probate forty-five days after writing; ERO, D/AMW 2/169 (1617), made probate thirty-one days after writing; and ERO, D/AMW 3/182 (1617), made probate sixty-eight days after writing. The handwriting in Robert Jacobb's will is not Thomas Jacob's and the scribe did not witness the document (or if he was one of the witnesses, he did not signal his role as scribe).

5. ERO, D/AMW 3/172.

my sonne" from Sir John Abatts. The amount of £55 had been repaid at the time the will was written, which must have been close to the time of Jacobb's death, as the probate date, when the terms of the document were deemed legally to have been fulfilled, was just forty-seven days after it was written.[6] Then there was another £100 that Jacobb owed Spencer, which was to be paid back at £4 per year on the final day of each subsequent November, with Humphrey's "admynistrators or assignes" being paid £50 within six months of Humphrey's death should it occur before the payment was completed.[7]

Having dealt with his debts to his brother, Jacobb then turned to his land and property, again following standard practice. A convention of many wills in this period was that the bequests were included in a particular order: bequests of land and property were usually included first, then money, followed by goods including, typically, furniture, bedding, tools, clothing, crops, and animals.[8] Not all wills follow this order, but many do; and not all wills include everything. Some can be very brief: "I bequeath all that I have to my wife" is a common single bequest in nuncupative wills.[9] Sometimes a glimpse is offered of a little more: in the 1628 nuncupative will of John Davison of Chadwell, for example, the witnesses—Richard Astley, his wife Anne, William Silkworth, and Elizabeth Davison, the dying man's wife—have had part of their recollections crossed through by the scribe. They remembered Davison saying "that all that he had he did give and bequeath to Elizabeth his wife saying that he was sorrye that he had noe more for her for that he should not

---

6. In the sample dataset of eighty-nine wills, 38 percent were made probate within sixty days and 64 percent within 180 days. Less than 20 percent of wills took over one year to be made probate. In this context, the forty-seven days it took for Robert Jacobb's will to be made probate would place it among the quickest 21 percent of the sample.

7. As with many wills from this period, planning for the deaths of one's legatees was necessary: the phrase "and if he should die before he reach the age of one and twenty I leave the said tenement to . . . " or variants on this, is a frequent refrain in wills. In an era when the average life expectancy was 35–40 years, leaving one's possessions to a younger generation was no guarantee that the bequest was secure.

8. Broadly speaking, this is how the exemplar wills in William West's *Symbolaeographia* are structured, although none of the four models he includes go into as much detail about goods as can be seen in the wills examined for this book. West, *Symbolaeographia. Which may be termed The Art, Description, or Image of Instrucments, Coueneants, Contracts, &c. OR The Notarie or Scriuener* (London: Richard Tothill, 1590), Lib. 3, Sect. 404–7.

9. See, for example, ERO, D/AEW 16/266 (1620), made probate within 123 days of speaking; ERO, D/ACW 9/95 (1622), made probate within fifty-seven days of speaking; ERO, D/ABW 49/42 (1629), made probate forty-three days after speaking. Some nuncupative wills lack a specific date on which the will was spoken and include a phrase such as "vpon a daye in November." In these instances I have taken the first day of the month named as the date of the will and thus can state that the will was made probate "within" a particular time.

~~leave her so well in estate as he found her~~"[10] Such struck-through sentiments tell us more about John Davison than does the general statement of his possession of things, but the scribe thought they were not appropriate to include in a legal document (although clearly Davison's witnesses who reported it thought otherwise).

While Elizabeth Davison inherited her husband's paltry estate, some women were left much better off—Jacobb, to return to our exemplary will, bequeathed his copyhold lands to his wife Johane "to haue and to hold . . . for and during the terme of her natural lieff if she shall so long live vnmaried."[11] The phrasing of his will is, once again, typical. It was the norm to leave a widow a right to dwell in her late husband's house "for her natural life," instead of giving her full legal possession of the property that she might then leave in her own will. Houses and land were customarily entailed in the late husband's will to be claimed by his (often, but not always, male) children after his widow's death. Thus, when property was bequeathed in a widow's (or even a wife's) will,[12] it is possible that that property may always already have been entailed to the legatee she names.[13] So when, in her 1628 will, Elizabeth Harvy noted that her

---

10. ERO, D/AEW 18/201 (1628), made probate fifteen days after speaking.

11. Andy Wood has argued that "generally speaking, copyhold tenure gave greater rights to women than did freehold and leasehold. . . . There were two means by which widows could inherit copyhold land. As widow's estate, regulated according to manorial custom, she might inherit a proportion (often a half, or more) of her dead husband's land. Alternatively, she might inherit a holding in the terms of her husband's will. This required the dying husband to name his wife as the inheritor of his lease. He could only do this if the copyhold was hereditable—that is, if it fell within the three lives' inheritance allowed in some manors; or under the stronger custom of copyhold by inheritance, in which copyholds were permanently hereditable." Wood, *The Memory of the People: Custom and Popular Senses of the Past in Early Modern England* (Cambridge: Cambridge University Press, 2013), pp. 298–99.

12. Those who listed themselves as wives in their wills (there are eleven in the dataset) often acknowledged their husband's permission to write a will. See, for example, ERO, D/ACW 9/110 (1622), made probate 125 days after writing; ERO, D/AEW 17/140 (1623), made probate 270 days after writing; ERO, D/ABW 47/271 (1625), made probate seventy-two days after writing; ERO, D/ABW 49/338 (1627), made probate eight days after writing; and ERO, D/ACW 11/268 (1632), made probate 213 days after writing. Alice Finch (al[ia]s Smith) noted in her will that "before his intermarriage with me" Richard Finch did "by his writing indented . . . covenant and graunt in consideracon of ye said marriage then to bee had . . . That it shall and may bee lawfull to and for mee . . . to make or declare my last will and testament, and in the said will to give ~~and~~ dispose, and bequeath fiftie pounds." ERO, D/AEW 19/218 (1633), made probate 5,029 days after writing. Some wives left bequests to sons who did not share the (living) husband's name, which suggests that the testators were remarried widows who were bequeathing their late husband's property to his children: see ERO, D/ABW 46/25 (1625), made probate 284 days after writing; and ERO, D/ABW 49/244 (1628), made probate 208 days after writing.

13. Although, as Carole Shammas notes, while married women were "*feme covert*, covered women, who no longer had any legal status," a woman still maintained her right to will or sell her own realty (i.e., property) if not personalty (goods). She goes on to note, however, that "because of the

late husband had bequeathed her "one house wth a garden, barne, stable and all other the apptinances therewithal . . . with the intent that I should give them all to wch of my twoe sonnes John or Samwell I should thinke fittinge," the fact that she was given the right to choose which son would inherit was unusual, but the preexisting bequest to the children was not.[14] Women's lesser status in relation to property is made evident more generally in the Essex dataset: out of the 885 wills by women (a figure that includes nuncupative wills[15]) only 162 (18 percent) include bequests of property of any kind. On the other hand, of the 3,559 wills by men in the dataset, 1,817 (51 percent) include property of some kind.[16]

To return to Jacobb's will, then, his bequeathing of his copyhold lands to his wife Johane "to haue and to hold . . . for and during the terme of her natural lieff if she shall so long live vnmaried" was orthodox. At her remarriage or death, "wch shall ffirst happen," the land would pass to Robert Jacobb's son, Robert the younger, "and to his heires and assignes fore ever."[17] Robert the younger also had to pay his mother £7 per year during her life, and to Humphrey Spencer

---

inheritance customs, women were less likely to have absolute rights over realty." Shammas, "English Inheritance Law and Its Transfer to the Colonies," *American Journal of Legal History* 31, no. 2 (1987): 147. In her will, Joan Hedge, a widow from Ramsey, for example, reiterated the bequests left in her late husband's will: she gave her elder daughter Martha Hedge "(beside the Ten pounds and ffifteene sheep wch her deceased ffather John Hedge gave her in his last will and Testament) all my wearing apparell linnen and woollen and my best payre of sheetes" and her son John Hedge "one sheepe beside the legacy his father gave him." ERO, D/ACW 11/138 (1632), made probate thirty-eight days after writing.

14. ERO, D/AEW 18/214 (1628), made probate 394 days after writing. A similar choice was offered to Joane, the wife of Robert Parker of Marks Tey, in his 1634 will, in which she was bequeathed all his movable goods and household stuff "and whateur else belonginge to the howse as also my Cattle being two Cowes & one Mare. Item my Corne in the Barne & my Crop vpō the grownd to be wholy hers to order & dispose where & upō whom she pleaseth at the day of her death if she dye a widowe: But if it shall seme good to the said Jone after my desease to contract & espouse her selfe in second Marriage: then my will & pleasure that my othe executor shall take all my Moveables to diuide equally betweene her & my . . . Cosen John Parker of Erles Colne." ERO, D/ACW 12/35 (1634), made probate 325 days after writing.

15. In the dataset overall, less than 8 percent of nuncupative wills include property, while 51 percent of all formal wills include it. Nine percent of men's nuncupative wills and 2 percent of women's nuncupative wills include property.

16. Even for widows, who make up 83 percent of the women in the dataset and who had legal rights that single women and wives did not have, only 18 percent of wills included land and/or property. Other women are listed as singlewoman (57), wife (13), daughter (1), spinster (26), maid (2), and "innholder and widow" (1). ERO, D/ABW 49/166 (1629), made probate 109 days after writing. Forty-seven wills list no status for the female testator.

17. An inversion of this can be found in the will of Mary Damyon, "wife," who bequeathed a copyhold tenement with all the appurtenances and three acres of ground adjoining to her husband Joseph Damyon "during the terme of his naturall life." After his death, the property was entailed to Elizabeth Howers "my sister her heirs executors adminres or Assignes for euer." ERO, D/ACW 12/33 (1634), made probate seventy-six days after writing.

"the full somes of £14 & £4 a yere" as specified earlier in the will. If the son failed to fulfill these conditions then his father's will was that the bequest of the land and property be made void and everything go instead to Humphrey Spencer and his heirs. Such conditions were not uncommon. The aims of wills of this time were typically conservative: they intended to maintain an order that had existed during the testator's life. Precluding the wife from holding the property if she remarried protected a bequest from incomers (the wife's next husband and her children with him), and ensuring that the son give his mother an income during her widowhood protected her from the son's possible neglect.

Jacobb senior next turned to his daughter Ann, to whom he bequeathed the sizeable sum of £100 to be paid to her by her brother six months after their mother's death (that is, when Robert the younger would have come in to full possession of the property). Her father also gave Ann the only animal in the will—"one heiffer the calf of a Cow called Goldelocks," perhaps as the start of her own herd. There followed two other small monetary bequests: to his "man" (that is, male servant) George 5s and to the female servant Susan 12d. Jacobb's will ends, as is conventional, with the bequest of what was left over— "I doe geue and bequeath all ^my^ goods and chattles whatsoer vnto the said Johane my Wieff & vnto him the said Robt my sonne."[18] Not only does this gather together all that remains, it also reinforces the nomination of executors, for, as Swinburne notes, "he to whom all or the residue is bequeathed, is thereby vnderstood to be made executor."[19] In his will, Jacobb made both his wife and son his executors, "but yet vppon this condicon not wth standing and my will and meaning is that my said wieff haue the sole vse and occupacon of all my said goods and chattels and shall stand as sole ~~ext~~ executrix of this my said will and testament vntill my sonne shall put in such security" for payment of his sister's legacy.[20] Again, the will was used to protect a legacy: the son's right of possession of the land was premised upon the daughter gaining her inheritance. This was not necessarily evidence of a dysfunctional family: such conditions were commonplace. Finally, Jacobb's will was signed by the testator, sealed, and witnessed by Arthur Hatch (who signed his name) and John Dames (who marked the will with a cross).[21]

---

18. ERO, D/AMW 3/172.

19. Swinburne, *A Briefe Treatise of Testaments and Last Willes*, p. 115r. Despite the masculine pronoun Swinburne used, 47 percent of wills in the sample dataset have female executors (wives account for 81 percent of those) and another 5.5 percent have both male and female executors.

20. ERO, D/AMW 3/172.

21. Literacy rates in the wills in the large ERO dataset echo data examined by David Cressy. Twenty-five percent of wills in the ERO dataset are signed; 75 percent are marked. There is a large

There is, thus, little that is unusual in Jacobb's will: the emphasis is on the valuable stuff he had accumulated—the land, house, and money—as is typical.[22] And apart from the two small bequests of money to servants, everything else—everything that was movable, which might include furniture, linen, clothing, tools, and animals, if other wills are a guide—was summed up in the words "goods and chattles"—except for Goldelocks's calf. The mention of the animals is uncommon: In the large ERO dataset, only just over 10 percent of wills include specific animal bequests and named animals are exceptionally rare (I return to the naming of animals in chapter 3). But Jacobb's lack of specificity concerning his other "goods and chattles" reflects the practice found in the majority of wills, with almost 60 percent of documents in the sample dataset also failing to offer any details about possessions beyond such a generic phrase. We should not take the lack of detail as evidence of lack of belongings, of course, and, in addition to having a house full of furniture, beds with bedding,[23] and, inevitably, clothing, I suggest that there would have been more animals in Jacobb's possession at the time he wrote his will than he included in it. This claim is significant in a book that will be using wills as a way of understanding people's relationships with their quick cattle and can be supported by an examination of Jacobb's status and by a comparison of wills with the inventories associated with them in the large ERO dataset.

---

gender distinction in the ability to sign: over 95 percent of all signed wills are by men. In his study of signatures and marks on the Protestation Oath of 1641, Cressy found that 63 percent of subscribers in Essex signed with a mark and that illiteracy rates for the sixteen parishes studied (1,081 subscribers in all) ranged from 85 to 36 percent. Cressy, *Literacy and the Social Order: Reading and Writing in Tudor and Stuart England* (Cambridge: Cambridge University Press, 1980), p. 73. We must be wary of assuming a simple link between signing a will and literacy, of course. Some testators were likely to be frail and so perhaps unable to sign their names because of illness even when they possessed the ability. Likewise, being able to read was not always linked with the ability to write: reading was taught before writing and some children only learned to read. Some of the wills might reflect this: in her 1625 will, the widow Joan Wood of Great Baddow, for example, bequeathed a Bible but signed with a mark. ERO, D/ABW 47/209 (1625), made probate fifty-five days after writing.

    22. I am using "stuff" here and throughout as my own catch-all phrase that signals all property—animal, vegetable, mineral, organic, manufactured. This was a common usage in the period covered in this book and deliberately avoids the connotations of terms such as cattle, chattel, and stock that are explored in chapter 1. *Oxford English Dictionary*, s.v. "stuff."

    23. These items would have been regarded as his rather than as property shared with his wife. On a number of occasions, the property the wife brought to the marriage was specified as her own. For example, in 1629, William Browne of Hatfield Peverel said: "Item I giue & bequeiue vnto my ^wife^ all the goodes she brought me." ERO, D/ACW 11/52 (1629), made probate sixteen days after writing. In the same year, a memorandum at the end of the will of Richard Colle, a yeoman from Ramsey, noted "that imediately after the finishing this the said Testator willed that Elizabeth his wife should haue all the goods houshould and Cattell which were hirs at the time of her marriage with him." ERO, D/ACW 11/42 (1629), made probate fourteen days after writing.

In seventeenth-century wills what appear to be designations of occupation can actually be designations of social class, so yeomen and husbandmen might, for example, participate in similar activities, but what marked them apart was their wealth, and how much land (rather than what kind of land—copyhold, leasehold, freehold) they owned was what made the crucial difference.[24] Those listed as laborers also participated in agricultural activity, but they worked for hire on the land of others, possessing only very small acreage for subsistence farming for themselves.[25] The meanings attached to the terms yeoman, husbandman, and laborer were thus very different from the meanings attached to the designations that reflected the particular trade of an individual—tailor, weaver, carpenter, shoemaker, and so on—although these latter individuals were also likely to be engaged in some kind of animal agricultural production, if only in order to support their own families.

While a testator's representation of their own occupation in a will might not always be beyond question,[26] Robert Jacobb's description of himself as a yeoman is supported by his landholdings: £255 would buy a great deal of real estate. Associated with his land would have been the inevitable possession of animals, for while Standon is in arable country in Hertfordshire, the river Rib runs through the village, meaning that meadow land would be likely to be available, and it is possible that it was some of this meadow that Jacobb had purchased.

---

24. Carole Shammas has suggested the following useful categories of what she terms "occupational status": "gentlemen (and all social ranks above); professionals (clergy, physicians etc.); merchants; yeomen . . . ; husbandman; craftsmen (weavers, shoemakers, small shopkeepers, etc.); and laborers (includes servants, apprentices, seamen, and soldiers)." She combines "spinster, wife, and widow into an occupational status category—women." Shammas, "The Determinants of Personal Wealth in Seventeenth-Century England and America," *Journal of Economic History* 37, no. 3 (1977): pp. 680–81.

25. In the dataset as a whole, 69 percent of gentlemen, 64 percent of yeomen, 37 percent of husbandmen, and 43 percent of laborers left property. It is often difficult to work out how much property was left by descriptions in wills, but the occupational status listed can also, it seems, reveal a gap between self-representation and reality when possessions listed exceed claimed status. For example, Nicholas Brande of Henham, who named himself a laborer in his 1621 will, bequeathed two properties: "The Tenemente wherin I nowe dwell with all & singular the howses yards Orchard & Chase waye thereunto belonging" and "the Tenement wherin my sonn Lewes Brande nowe dwelleth with the yards Orchards gardins Pasture ground or meddowe & Chase waye." ERO, D/ACW 9/26 (1621), made probate thirteen days after writing. Thomas Rudland, a laborer of Colchester, left one "ffreehold message and tennenent," and also bequeathed the rent from "my other message and tenement." ERO, D/ACW 11/27 (1629), made probate forty-six days after speaking. On the other hand, the laborer Nicholas Parnell of Elsenham left only "the little house wherein I now dwell ^with the orchard in the back field^." ERO, D/ACW 10/128 (1625) made probate 686 days after writing). And William Hodgekin of Hadstock, another laborer, bequeathed his son "the lease and chattel I haue for certayne yeres to come of hopground ^and hoppoles wth [illeg.]^ meadow pasture and lays and wood . . . being from John Adams . . . yeoman." ERO, D/ACW 12/36 (1634), made probate thirty days after writing.

26. In the formal will and appended nuncupative will of William Pease of Great Baddow, for example, we can see a gap between self-representation and socially understood status. In his formal will (dated 17 July 1622), Pease defined himself as a yeoman, whereas in the nuncupative will that

However, even if Jacobb possessed only arable land, it is probable that he would also have had a number of cows for the household's dairy; milk, in particular for cheese and butter making, was central to the early modern diet.[27]

Another will from Standon from the year after Jacobb's offers a useful comparison. The widow Joane Meritowne includes three separate bequests of land in the parish in hers. These bequests are all to her daughters: she left "one acre of arable land" and another "acre & a rode more or lesse" to Mary; two acres of arable to daughter Joane, with the "tenemts wth ^all the^ yards garden and orchard wth all & singular thapptenncs" in the same parish; and another acre of arable land to daughter Grace. Along with the arable land, the will also included two cows: her son Thomas got "my best cowe & one payre of flexen sheats" (perhaps Thomas had already received a bequest of land from his mother during her lifetime, hence his lack of such a bequest here).[28] In addition, "Henry Clark my man" got one cow "for his good service." Meritowne was apparently less well off than Jacobb, for while she was in possession of land it was in small parcels. But the will also contained a number of monetary bequests that might signal something else: one "tenant" got 10s, and two others were forgiven the rent they owed her. This inclusion of references to tenants implies that she may have owned more land than she bequeathed to her daughters—although, of course, their bequests may be of the land that brought in a rental income.[29] Meritowne's will also tells us that she had at least three cows: there was a "best cowe" and "one cow" (not "the other cow"), so there was likely to have been at least one more. A comparison with the will of another widow, Elizabeth Cooke of Great Braxted in Essex, reinforces this sense of the number of animals in wills. Cooke bequeathed to

---

makes up a codicil (dated July 1623), his neighbors described him as a saddler. ERO, D/ABW 45/92 (1623), made probate 342 days after writing. On this issue, see Margaret Spufford, "The Limitations of the Probate Inventory," in *English Rural Society, 1500–1800: Essays in Honour of Joan Thirsk*, edited by John Chartres & David Hey (Cambridge: Cambridge University Press, 1990), p. 144.

27. See Craig Muldrew, *Food, Energy and the Creation of Industriousness: Work and Material Culture in Agrarian England, 1550–1780* (Cambridge: Cambridge University Press, 2011).

28. ERO, D/AMW 3/157 (1618), made probate twenty-three days after writing. In his will of 1623, for example, the yeoman Peter Hove of Southweald revealed that he had already begun the process of bequeathing land during his life. He noted that he had "already caused his son Robert to become bound by sev'all Obligacons" to his siblings "for payment of severall somes of money vnto them . . . and in Consideracon thereof haue surrendred my Coppyhold lands and Tenemente called Paynes and ffernecrofte to the vse of my said sonne Robert and his heires." ERO, D/AEW 17/145 (1623), made probate 148 days after writing. Likewise, Richard Lake of Theydon Garnon's will included the following: "Whereas my daughter Elizabeth receiued a portion vpon hir marriage I hould it not fit to giue her an equall portion with the rest of my children and therefore now giue hir onely ten poundes." ERO, D/AEW 19/269 (1634), made probate 3,063 days after writing.

29. ERO, D/AMW 3/157.

Jane her daughter "a Cowe wch shee shall choose," to another daughter Alice Rawlins "my seconde Cowe," and to her son John "the other Cowe."[30] In the light of evidence such as this of Meritowne's possession of cows, Jacobb, with (at least) £255 worth of land, was likely to have had more animals than just Goldelocks and her calf. But they, of course, are all that are referred to in the will. However, failure to include all livestock in this document is not particular to Jacobb. Indeed, of the seven yeomen from Standon who had wills made probate in the period 1615 to 1619 in the ERO, Jacobb is the only one who included specific animal bequests.[31]

In addition to the evidence from contemporary wills made by possible neighbors, an analysis of inventories from the period reinforces the likelihood that Jacobb possessed more animals than he specified in his will. Inventories were the lists and valuation of property (excluding land and "grasse or trees" which pass with the land to the heir) that were taken shortly after an individual's death by two or more neighbors.[32] These documents reveal that despite the fact that animals might be only rarely specified as bequests in wills, they were very often in an individual's possession; and thus that lack of livestock in a will does not automatically signal that the testator was not involved in animal agriculture. Inventories in the Essex Record Office from the period of the large ERO dataset are rare, but a few have survived and help make this point.

The nuncupative will of Richard Cesar of Terling includes an inventory of his goods. In his testament of 2 December 1622, given just five days before he was buried in All Saints church, Cesar left his son Richard 40s "and an old chest," he left his other son Henry "one little hogge," and he left his (nameless) wife (to whom we return in chapter 1) "all the rest of my goods moveable

---

30. ERO, D/ACW 11/197 (1631), made probate 132 days after writing.

31. In his will written three months after Jacobb's, Nicholas Rudde bequeathed "all my freeland" to his eldest son, a "Tenement or Cottage with a backside" to each of his two other sons, and monetary bequests totaling £150, including £20 to each of his five daughters and another £20 to "the Child which my wife now goeth with, whither it be sonne or daughter." The document made no reference beyond the catchall phrase "all the rest of my goods Cattell Chattels & Implements within the house or without vnbequeathed" to livestock. ERO, D/AMW 1/144 (1618), made probate twenty-six days after writing. Likewise, in his will written in July 1615 and probated in April 1617, William Barfoote, another yeoman from Standon, bequeathed his younger son Nicholas £60, his elder son William two closes of land totaling twelve acres, and his daughter Winifrid "that dwelling howse ^& the wod house^ wth the yard that shee now dwell in" with "free ingresse, egresse & regresse to the ponde in the meade by the said howse to fetch water as her neede shall require." Two granddaughters, Elizabeth and Margaret, each got £10, and his grandson Thomas got 6s 8d, perhaps because at a later date he would be inheriting the land his father William had received. No animals were mentioned. ERO, D/AMW 1/72 (1617), made probate 649 days after writing.

32. Swinburne, *A Briefe Treatise of Testaments and Last Willes*, p. 218v.

and vnmoveable paying the Summe of five and thirty shillings fower pence which I owe [and] discharging all my ffunerall expenses."[33] On the same sheet of paper, and in the same hand, is the "true Inventory" of Cesar's goods, which William Drayne and William Gotch took on 6 February 1623. The inventory includes five items only: "two bullockes" valued at 53s 4d; "his shoppe tooles," worth 26s 8d; "one little hogge," valued at 3s 4d; "one little chest," valued at 1s 4d; and "his purse and apparell," valued at 5s. It is likely that the unusual amount of time that had elapsed between Cesar's death (which took place sometime between 2 and 7 December) and the taking of the inventory two months later might have impacted what possessions were listed, with the distribution of property perhaps already under way, but if we set aside Cesar's debts of 35s 4d, the total value of his estate as represented in the inventory was £4 9s 8d, which means that the two bullocks accounted for over 59 percent of the overall value of his possessions and that all of his animals accounted for over 63 percent. However, by worth, less than 6 percent of the animals (i.e., the pig) were present in the will itself.

Another nuncupative will that includes an inventory is that of William Younge, a weaver from Roydon, dated 13 January 1628, three days before his burial in St Peter's and eleven days before the inventory was taken.[34] The document is unusual in two ways other than its inclusion of an inventory. First, it includes another named cow, and second, it is written in the style of a formal will, including the opening statements that are missing from the majority of nuncupative wills. However, after the preamble, the will includes only three bequests, which makes it more like a conventional nuncupative will: Younge left 40s to his unnamed sister (the scribe, who had clearly forgotten her name, left a gap in the text to put it in later, but did not go back to do so[35]); a heifer called Nan to his "nursegirl" Katherine Anger; and "all the rest of my goods chattell, cattell & moueables" to Kathrine, his wife.[36] On the reverse side of the will and written in the same hand, the inventory gives the detail and monetary value of all the "goods chattell, cattell &

---

33. ERO, D/ACW 9/62 (1622), made probate sixty-six days after speaking. Cesar's burial (where he is listed as Richard "Sizer") is recorded in the Parish Register of All Saints, Terling, ERO, D/P 299/1/3 (1538–1688). The fact that Cesar and Sysaye (the spelling of the family name given in Cesar's widow's will) are the same was brought to my attention by Wrightson and Levine in Poverty and Piety in an English Village, p. 39.

34. His burial is recorded in the Parish Register of St Peter, Roydon, ERO, D/P 60/1/1 (1567–1706).

35. As Henry Swinburne notes, such errors on the part of the scribe "ought not to preuaile against the truth of the testament." Swinburne, A Briefe Treatise of Testaments and Last Willes, p. 190v.

36. ERO, D/AMW 2/125 (1628), made probate forty-five days after speaking.

moueables" Younge included in his bequest to his wife. The list is typical of the period: there is furniture, bedding, kitchen equipment, working tools, animals, money, and clothing. After the £20 in "ready money" (which makes up more than 31 percent of the overall value of his possessions), the animals—a mare, three "bease" (cows), a bullock, a hog, three sheep, and some poultry—are the most valuable things in the will.[37] Together they make up nearly 19 percent of the inventory's overall value, but only one of these creatures—Nan—appears in the will.

Another fourteen inventories with associated wills can be found in the large ERO dataset,[38] and of the sixteen surviving inventories, thirteen (81 percent) include animals, while only five of the sixteen associated wills (31 percent) include specific animal bequests, with one remaining unclear in its summarizing statement: "All the rest of my goods cattells chattels move-ables vtensills iewells."[39] The 31 percent of wills that include specific animal bequests in this group with associated inventories is a much higher propor-tion than in the large ERO dataset, where it is 10.1 percent. There is no logical reason for the higher proportion of animals in wills in the inventory group. Over half—56 percent—of the wills with inventories are nuncupative, which is likewise a much larger proportion than can be found in the large ERO dataset (where they make up 14 percent of all wills). The abundance of nuncupative wills in this smaller subsample should actually push the per-centage of animals specified in the wills down, because in general, while 10.1 percent of all wills include specific animal bequests, only 7 percent of nuncupative wills did so. It is, perhaps, nothing more than chance that a high proportion of wills with particular animal bequests have survived with their inventories in the ERO.

Despite this anomalous aspect of the wills with inventories, the proportion of those inventories that include animals from the period of the large ERO data-set (over 80 percent) is not totally at odds with what can be tracked in other con-temporary or near-contemporary ones. For example, fifteen (68 percent) of the small sample of twenty-two Essex inventories from 1635 to 1640 in F. W. Steer's

---

37. ERO, D/AMW 2/125.

38. Three other wills have inventories with them, but these were made by the testator at the time of drawing up the will and did not include financial valuations. See ERO, D/AMW 4/41 (1629), made probate ninety-three days after writing; ERO, D/ABW 50/65 (1630), made probate 1,166 days after writing; and ERO, D/ABW 52/170 (1634), made probate forty days after writing.

39. ERO, D/APgW 1/17 (1632), made probate 151 days after writing.

Farm and Cottage Inventories include animals.[40] In a study of wills and inventories from Stanton-by-Bridge, Derbyshire, D. J. Baker found that "nearly all the inventories before 1600 showed the value of the animals making up half or more of the total estate"; and Carole Shammas likewise underlines how commonplace animal ownership was in the first half of the seventeenth century: of the seventy-three inventories she analyzed from Oxfordshire and Northamptonshire for the period 1600–1649, 53 percent included cows, 47 percent included pigs, 38 percent included sheep, and 29 percent included poultry.[41]

In their much larger study of 8,103 inventories, Mark Overton, Jane Whittle, Darron Dean, and Andrew Hann found that in Kent in the period 1600–1629, 55 percent contained evidence of "potential" dairy production for use within the household; the comparable figure for the period 1630–1639 was 56 percent. Dairy production means cows as well as the equipment that would be used in milking and to make butter and cheese. These figures therefore do not include the presence of other livestock animals (sheep and pigs, for example).[42] And, as a final example here, in his study of 972 inventories of the period 1550–1800, Craig Muldrew found that 68 percent of laborers owned animals.[43] Thus, on the basis of the evidence from extant contemporary or near-contemporary inventories, as from the evidence of geography and

---

40. F. W. Steer, *Farm and Cottage Inventories of Mid-Essex 1635-1749* (Essex C.C., Colchester: Wiles and Son, 1950). Steer's collection of Essex inventories begins as my dataset ends, marking when the survival rate of those documents in the county goes up. Even when using the figures that extant inventories suggest, it is worth remembering the limitations of these documents. These are explored in Lena Cowen Orlin, "Fictions of the Early Modern English Probate Inventory," in *The Culture of Capital: Property, Cities, and Knowledge in Early Modern England*, edited by Henry S. Turner (New York and London: Routledge, 2002), pp. 51–83.

41. D. J. Baker, "Stanton-by-Bridge: A Study of Its People from Wills and Inventories, 1537–1755," *Derbyshire Miscellany* 10, no. 3 (1984): 81. See also Carole Shammas, *The Pre-Industrial Consumer in England and America* (Oxford: Clarendon Press, 1990), p. 21. Shammas's figures don't include the ownership of horses.

42. According to their study, the proportion of inventories that included evidence of potential dairy production found in Cornwall was slightly higher than the proportion in Kent: 60 percent for 1600–1629 and 59 percent for 1630–1659. Mark Overton, Jane Whittle, Darron Dean, and Andrew Hann, *Production and Consumption in English Households, 1600–1750* (London: Routledge, 2012), p. 57.

43. Muldrew, *Food, Energy and the Creation of Industriousness*, p. 166. Figures from probate inventories from Northamptonshire from the first half of the eighteenth century show similar proportions: 57 percent of laborers and 97 percent of farmers owned cattle, 23 percent of laborers and 79 percent of farmers owned pigs, and 25 percent of laborers and 74 percent of farmers owned sheep. Leigh Shaw-Taylor, "Chapter Three: The Nature and Scale of the Cottage Economy," table 3.3, Occupations Project Paper no. 15, Cambridge Group for the Study of Population and Social Structure, http://www.geog.cam.ac.uk/research/projects/occupations/abstracts/paper15.pdf, accessed 21 July 2016.

status, we can suppose that Robert Jacobb's inclusion of one specific animal bequest in his will may not reflect the full extent of his livestock holdings. There are other members of the herd waiting in the wings, gathered into the phrase "all ^my^ goods and chattles whatsoer."

More evidence of the fact that animals were present but not mentioned in wills can be found in a record of tithe payments. In 1627, Charles Chadwick, the vicar of Woodham Ferrers, recorded in his commonplace book the animal property and the related tithe payments of a number of his parishioners. The wills of three of those parishioners can be traced, and while Richard Linne was recorded as having paid "for 4 Calues & one lamb 7s 6d," Henry Cooch paid for "4: Calues a lamb & a halfe 20s," and George Drywood paid "for his tythe 1626: 5: Calues 5s & 7 lambes 2s 8d," none of these men included live-stock in their wills.[44]

Given that most testators seem not to have included animals in the posses-sions they bequeathed in their wills, we might wonder what it is that wills can offer to our understanding of human-livestock relations. It might appear that in addition to offering a skewed picture of the population because so few were made by those in the lowest social groups, these documents might reveal at worst only the absence of animals and at best their constant presence as simply another kind of temporal possession, included only under a general catch-all term. It is inevitable that this might be the case: on one level, all ani-mals in wills are just temporal possessions with financial value. That is, after all, what wills deal in. But on the other hand, as the chapters that follow aim to show, the descriptions of the animals in wills often reveal much more than that. I will suggest in chapters that follow that naming is evidence that these animals might have had more than a financial meaning to the testator, as do some of the very careful descriptions that are present in some wills. It is here, in the realm of qualitative rather than quantitative analysis, that wills can become useful sources for an understanding of human-livestock relations in a period when so many of those who lived with and worked with animals were illiterate and so many of the day-to-day practices they engaged in were never

---

44. The wills are (respectively) ERO, D/ABW 55/248 (1639), made probate 307 days after writing; ERO, D/ABW 53/166 (1636), made probate eighteen days after writing; and ERO, D/AEW 20/179 (1638), made probate seventy-eight days after writing. Drywood's will is classed differently from his neighbors' as he is listed as living in Mundon (five miles away and in another archdiocese), but I am assuming that this is the same George Drywood as Chadwick records because in his will he requests burial in Woodham Ferrers. Chadwick's Common Place Book is ERO, D/Dra Z1 (1611–1627).

written down. In addition to this detail about otherwise invisible lives (both human and animal), certain family relationships can be traced in new ways through tracking the bequeathing of livestock. Animals were not only shared, they were also homed with others and lent out. In addition, who received what animals—the frequent giving of lambs and other young animals to children, for example (a focus of chapter 4)—might tell us something about all parties to the bequest. It is these aspects of early modern culture that the wills in Essex offer glimpses of.

Such animals as can be tracked in the wills have too often been written out of the record or have been written in only as objects of use. This is as true of writings from the early modern period as it is of writings about the early modern period. Thus, discussions of animal rationality in the seventeenth century are often purely theoretical, and real animals—animals who bark, moo, neigh, kick, bite, smell—and those who work with them tend to be absent.[45] When pondering whether dogs could reason in a public debate staged for the king in 1615, for example, scholars from Cambridge University were happy to invoke the ancient philosopher Chrysippus as their evidence, but it was the great huntsman James VI and I who interrupted their discussion by referring to his own (real) pack of hounds. Likewise, where René Descartes' beast-machine hypothesis denied thought to animals and regarded them as automata and ignored the differences between a monkey and a sponge, William Cavendish, the Earl of Newcastle, wondered in response how he could train a horse if the horse was not able to form "a judgment by what is past of what is to come (which . . . is thought)."[46] Although the philosophers could, it seemed, write their philosophy without real animals, their critics often positioned themselves as those who lived with and worked with animals, and those critics might question the philosophers simply by bringing those real animals into the discussion.

In a strikingly similar way, animals are frequently reduced to little more than the machines of (human) history in recent studies of the past. Undoubtedly important work in agricultural history, for example, has shown the changing weights of animals across time, the growing herd sizes, and so on.[47]

---

45. I see this as a difference between the view from on horseback and the view while walking alongside the animal. See Erica Fudge, "The Human Face of Early Modern England," *Angelaki* 16, no. 1 (2011): 97–110.

46. Both are discussed in Erica Fudge, *Brutal Reasoning: Animals, Rationality and Humanity in Early Modern England* (Ithaca, NY: Cornell University Press, 2006), pp. 101–4, 163–64.

47. I am, of course, indebted to work from this field in what follows, as footnotes will reveal. My point is that agricultural history has not taken up the issue of animals as sentient creatures and that this adds a new focus to the study of agriculture.

Lost in these figures, however, are the living, individual animals—the animals with whom people interacted, often daily. This book attempts to put those real animals back into the picture; it attempts to bring into focus the fact that, for example, the plowman could not be a plowman without his team of horses and that the team, perhaps, was better thought of as a human-horse collaboration.[48] The book will try to give meaning to the claim that, according to Leonard Mascall's 1627 figure, "a Cow will liue well fifteene yeares, but after that she will ware feeble & weary."[49] What this might mean for the woman who milked the cow—leaned against her flank, twice a day, day in, day out, for over a decade—and for the cow, too, is worth exploring if we are to attempt to understand the period in which that relationship was so commonplace and so vital. Other creatures—pigs, sheep, chickens, and bees—will also play their roles here (and one civet cat will raise her lonely head), but it is the people who give us the lead. However sparse, however limited the world they represent might be, their wills—often the only documents these people left us—offer an insight into their lives. And those lives, I suggest, cannot be comprehended without including the animals.

I said at the beginning of this introduction that there was one cow who could not get into this book. That, of course, is not true: Goldelocks is in the book. She is the first individual animal we have encountered. But she is outside the book in another way. Histories are written where records allow, and Essex has proved fertile territory for numerous important studies of early modern culture that form a backbone for my own work. With boundaries within about five miles of the City of London; borders with the counties of Cambridgeshire, Hertfordshire, Middlesex, and Suffolk; and a substantial coastline on the North Sea, Essex provides historians with a range of economic and environmental possibilities. These include enclosed and common field systems, new trades and agricultural labor, coastal and inland regions, cities, towns and small villages. John Norden depicted the county in 1594 as "moste fatt, frutefull, and full of profitable thinges, exceeding (as farr as I can finde) anie other shire, for the general comodities, and the plentie."[50] And in the 1610 English language edition of William Camden's *Britannia*, it was

---

48. Of the one plowman and eight plowwrights in the database, none include animals in their wills: one plowwright referred to "my stocke or goods." ERO, D/ABW 48/209 (1626), made probate 387 days after writing.

49. Leonard Mascall, *The Gouernment of Cattell* (London: Tho. Purfoot, 1627), p. 53.

50. John Norden, *Speculi Britanniae Pars: An Historical and Chorographical Description of the County of Essex* (1594; repr., London: Camden Society, 1840), p. 7.

likewise described as being "large in Compasse, fruitfull, full of Woods, plentifull of Saffron, and very wealthy; encircled, as it were, on the one side with the maine sea, on the other with fishfull Rivers which also doe affoord their peculiar commodities in great abundance."[51] The lives of the circa 100,000 people who inhabited the county in the first half of the seventeenth century thus offer glimpses of worlds of small-scale subsistence, dual employment, and the growing size and economic significance of the capital in the expansion of a yeoman class. The county also boasts an exceptional archive, and studies of Essex that have used these records have offered crucial insight into social, criminal, and family lives from the period.[52] In these studies, the worlds of those who lived in early modern Essex are brought back into view and I hope that this study will add to our understanding. In addition to the records from Essex and to the histories written with those records, studies of early modern culture that take other places as their focus are important critical contexts here. And while histories have shown how far geographical specificity might form local culture and so suggest that generalizations from detailed archival studies must be resisted (what is found in Cambridgeshire might not be found in Essex, for example), it is inevitable that what is discovered in one region will influence what is at least looked for in another.[53] For more than the dictates of bureaucracy (where records are kept), then, territorial boundaries are often used to limit studies. My territory is Essex, and, sadly, Goldelocks lived over the county boundary in Hertfordshire, even though the will that refers to her is included in the Essex Record Office—because the wills were organized by diocese rather than by county: Jacobb's will is from the Archdeaconry of Middlesex (Essex and Hertfordshire Jurisdiction). For this reason, she is not part of the data here.

---

51. William Camden, *Britain, or A chorographicall description of the most flourishing kingdomes, England, Scotland and Ireland*, translated by Philemon Holland (London: Iohn Norton, 1610), p. 439. See also Joan Thirsk, "Farming Regions of England," in *The Agrarian History of England and Wales*, vol. 4, *1500–1640*, edited by Joan Thirsk (Cambridge: Cambridge University Press, 1967), pp. 53–54.

52. See, for example, Alan Macfarlane, *The Family Life of Ralph Josselin: An Essay in Historical Anthropology* (Cambridge: Cambridge University Press, 1970); Alan Macfarlane, *Marriage and Love in England: Modes of Reproduction, 1300–1840* (Oxford: Basil Blackwell, 1986); J. A. Sharpe, *Crime in Seventeenth-Century England: A County Study* (Cambridge: Cambridge University Press, 1983); and Wrightson and Levine, *Poverty and Piety in an English Village*.

53. See, for example, Joan Thirsk's argument about the impact of enclosure on the social lives of what had been common field villages; and David Underdown's reading that the nature of the agricultural region had an effect on political allegiance in the English Civil War; Joan Thirsk, "Enclosing and Engrossing," in Thirsk, *The Agrarian History of England*, p. 255; David Underdown, *Revel, Riot, and Rebellion: Popular Politics and Culture in England, 1603–1660* (1985; repr., Oxford: Oxford University Press, 1991), pp. 40–41.

Another limiting factor I have put in place excludes Goldelocks too. Any piece of historical research has to draw its own temporal boundaries. Sometimes those boundaries are obvious: a study of Tudor England or the history of a specific monarch's reign, for example. But temporal boundaries might not be organized around monarchs and their lineage. The classic *Agrarian History of England and Wales* divides the past differently: Volume IV covers 1500–1640 and Volume V covers 1640–1750. Many more examples could be offered, and my point is not to debate these differences in periodization but to recognize that the focus of the history underpins how the researcher chooses the beginning and end points of a time period.[54] In selecting a period for the focus of my archival work on the study of human-livestock relations in early modern England, I made a deliberate choice to avoid the dates of monarchs—dates which would have meant nothing to the animals. Rather, I opted for a period with practical human as well bovine considerations: I chose one in which there was a good survival rate of wills and one that mirrored the potential productive life of an early modern dairy cow. My period is fifteen years long—1 January 1620 to 31 December 1634: after that I, too, ware feeble & weary. The narrow nature of this focus allows for a detailed glimpse of a short moment but cannot reveal significant shifts over a long period of time, but in offering the detail, I hope to outline evidence for lives that have thus far been excluded from historical analysis. Goldelocks comes too early to be included here: Robert Jacobb's will is dated 1617. But we can still learn much from her.

After all the detail of the debt he owed his brother and how he planned to repay it, after outlining what would happen to his wife after his death and his son's responsibility to her, and after guaranteeing how the sizeable financial legacy left to his daughter would be paid by linking it to his son's own inheritance, Robert Jacobb made only four other bequests in his will: two small monetary gifts that totaled 5s 12d, Goldelocks's calf, and everything else. I began this book with Goldelocks because she is emblematic of the problem we encounter when we attempt to think about relationships between people and quick cattle in early modern England. She is there, but she is also absent. Not only have my arbitrary choices of the location and dates excluded her, but Jacobb's representation of her also—paradoxically—pushes her to the margins. It is not the named cow, Goldelocks, but her anonymous calf who

---

54. It is now well established that the periodization of western history is not meaningful outside the west. See, for example, Henk Wesseling, "Overseas History," in *New Perspectives on Historical Writing*, edited by Peter Burke (Oxford: Blackwell, 1991), pp. 67–92.

is important in Jacobb's will; it is the heifer who is the bequest. Goldelocks, even as she is referred to, even as she is the one with the name, is in the will simply as the instrument for the production and identification of the valuable legacy. But therein lies her worth to Jacobb. There are a number of ways of viewing this: if Goldelocks was not bequeathed because she was already dead when Jacobb came to write his will, then her presence as the means by which a particular heifer might be identified signals that she was a recognizable individual or perhaps even that she was a cow who produced particularly high-quality offspring. If the latter is the case, recognizing a heifer as coming from this particular dam would be significant for the family's future breeding plans. But if Goldelocks was alive at the time Jacobb wrote his will, it would be his wife Johane and his son Robert who would inherit this animal (she, like all the other possible cows in her herd, was simply part of the goods and chattels gathered up toward the end of the will). But even with this (non)status, we must assume that if she was alive, Robert Jacobb the testator assumed that Goldelocks would keep on producing milk and calves (nutrition and wealth), that she would continue to be crucial to the well-being of the family he was about to leave behind. Goldelocks thus had a name, she was not quite present, she was simply an instrument, and she was essential. All of these things were happening simultaneously.

This is the complexity of human-animal relations that I hope to begin to unpack in the five chapters that follow. In chapter 1, I look at the absence of animals in almost 90 percent of wills in the large ERO dataset and read the collecting phrase at the end of wills in the context of discussions of will writing in the period. What might seem to be careless generalization emerges, I propose, as sensible future planning. Chapter 2 uses the specific animal bequests in just over 10 percent of wills to trace what life with quick cattle would have been like for many, using agricultural manuals from the period together with the wills to add to our understanding of labor alongside animals. Chapter 3 gets closer still to the animals and takes its lead from wills that include animals—cows—with names. These are rare, but further discussion of the need for human-cow collaboration in the practice of milking offers new insight into interspecies relationships in the period. Chapter 4 turns to bees and lambs to think about the possibility that some specific animal bequests, and thus some animals, had meaning that went beyond the financial. Symbolic systems of the period will be read as being linked to animal legacies but also to husbandry practices. Finally, chapter 5 leaves Essex to assess the impact of urbanization—the growth of London—on human relationships with animals in this period. Additional wills—from the Diocese of London and a few from the Prerogative Court of Canterbury—are placed alongside the large

ERO dataset and a picture emerges of a very different world coming into being. The afterword turns from the early modern period to think about what the historical work might add to current discussions of farm animal welfare, and to an understanding of the industrial agriculture of the twenty-first century. While the focus of the book is on the early seventeenth century, I hope that Goldelocks and the other quick cattle we encounter might also offer new ways of thinking about where we are now and where we might go from here.

In the second half of the sixteenth century, the great agricultural author Thomas Tusser wrote "Thy soule hath a clog. Forget not thy dog."[55] For Tusser, the responsibility of the early modern husbandman and housewife did not stop with the human household: it included all the animals on the land too. Writing histories of those people without paying attention to the animals is thus, to use Tusser's phrasing, a kind of forgetting.[56] I hope that what follows might begin to unclog an aspect of the past that has until now been largely obstructed, that it might free some ideas from the blockage of silence and begin to show what attention to early modern livestock and the worlds they shared with humans can bring to our understanding of the past and the present.

---

55. Thomas Tusser, *Five Hundred points of good Husbandry* (London: I. O., 1638), p. 133. This is advice to the housewife to remember to feed her "bandog that serueth for diuers mishaps." Andrew McRae notes that "Tusser's husbandry manual went through twenty-three editions in eighty-one years, after its first publication in 1557 as *A hundreth good pointes of husbandrie*." McRae, *God Speed the Plough: The Representation of Agrarian England, 1500–1660* (Cambridge: Cambridge University Press, 1996), p. 5.

56. I am reminded of what Jacques Derrida termed a "calculated forgetting"—a forgetting that underpins, and allows for, the kind of history we have. Derrida wrote: "And from the vantage of this being-there-before-me [the animal] can allow itself to be looked at, no doubt, but also—something that philosophy perhaps forgets, perhaps being this calculated forgetting itself—it can look at me. It has a point of view regarding me." Jacques Derrida, *The Animal That Therefore I Am*, edited by Marie-Louise Mallet, translated by David Wills (New York: Fordham University Press, 2008), p. 11.

❧ CHAPTER 1

Counting Chickens in Early Modern Essex

Writing Animals into Early Modern Wills

In a nuncupative will dated 24 April 1625, that has an associated inventory, Widow Sysaye of Terling (the widow of Richard Cesar, mentioned in the introduction), did what is by now familiar: she did not include all of her belongings. In the statement given in front of witnesses who included William Gotch (the same man who had taken Cesar's inventory a few years earlier), she bequeathed 80s to the parish for the "puttinge foorth" of her two sons, the "little boy" Henry and the "bigger boy" Richard. In addition, she left her unnamed daughter a sheet, a deep kettle, two platters, "a forme [that is, a bench] and, a table which her daughter bought with her owne money," one green apron, a hat, and a gown. The widow's unnamed son, "which she had by her first husbande," received 4d, and Mary Gotch, William's wife, got 19s 6d "for the charges at her burial" with the directions that she should "deliuer the rest [i.e., what was left over] to the parishe."[1]

The inventory associated with the will, taken by Jeremy Cuson, John Clarke, and John Drayne just four days after the nuncupative statement was made and two days before the widow's burial, reveals the wider extent of

---

1. ERO, D/ACW 10/115 (1625), made probate 134 days after speaking. It is possible that Widow Sysaye's unnamed son had already received his inheritance from his late father and that his mother's bequest of 4d was simply to mark that she had not forgotten him.

her possessions.[2] It includes six pieces of pewter, some candlesticks, two beds with bedding, a cupboard "and woole in it," her clothes, linen, kitchen equipment, and her animals. These were (in the order they are given in the inventory) two hens valued at 20d (the first item on the inventory), a hog valued at 8s (the 22nd item), and at the bottom of the list "a henne and ten chickens" valued at 2s, "5 chickens" (2s 6d), and, on the final line, "2 more" (6d).[3]

What I am interested in here is not the fact of the presence of the animals in the inventory of Widow Sysaye's possessions: rather, I am interested in the inventory's unusual representation of the hens and chicks.[4] Other inventories from the period of the large ERO dataset used more general collective terminology to describe the birds. William Younge's inventory used the single word "poultry," giving it the value of 6s 8d, the inventory of Edmund Ellis of Good Easter used the same term, "poultrie," and valued it 5s, the inventory of Thomas Eve, also of Good Easter, mentioned "poultery in the yard" worth 13s 4d, and the inventory of another Good Easter resident, William Cooche, referred to "fowles in the yard" worth 7s.[5] Widow Sysaye's inventory is rather different. The first item in the list of her poultry is her two hens, and then later in the inventory the assessors listed a hen and her ten chicks. Then there are another five chicks. Then two more. It is as if the way the inventory is written reenacts the neighbors' problems counting the birds as they moved about the yard. There are some. There are some more. Oh, and some more. This is not so much the difficulty of counting chickens before they hatch, but simply the difficulty of counting chickens.

Widow Sysaye's birds offer a material illustration of what was a more abstract problem for will writers: how to include animals. Individualization through detailed description will be discussed in the following chapter, and the implications of naming animals is the focus of the chapter after that, but this chapter looks at the terminology (legal and otherwise) people used to gather up what were constantly moving, changing beings into clear, concise, and perhaps we might call them "clean" categories. It analyzes how what might

---

2. "Priscilla Sizer, widow" was buried on 30 April 1625. Parish Register of All Saints, Terling, ERO, D/P 299/1/3 (1538–1688).

3. ERO, D/ACW 10/115.

4. The word chicken was used to describe only young birds until the nineteenth century, when it began to be used to describe "a domestic fowl of any age." *Oxford English Dictionary*, s.v. "chicken."

5. ERO, D/AMW 2/125 (1628), made probate 45 days after writing; ERO, D/APgW 1/15 (1626), made probate 136 days after writing; ERO, D/APgW 1/8 (1621), made probate 1,627 days after writing; ERO, D/APgW 1/17 (1632), made probate 151 days after writing. Seven of the small group of nine wills from Good Easter (classmark D/APgW) in the large ERO dataset include inventories.

be scattered over the fields and yards was tidied up in the written document and asks whether clarity and succinctness is, and was viewed as, the best way to transfer all possessions. John Law, a sociologist of science and technology, offers a valuable way of addressing the question of clarity. He proposes that the attempt to "describe things that are complex, diffuse and messy . . . tends to make a mess of it . . . because simple clear descriptions don't work if what they are describing is not itself very coherent." His focus is on how studies in the social sciences (mis)represent the world, but his ideas can, I think, be usefully transferred to think about writing wills in early modern England. Law continues: "If much of reality is ephemeral and elusive, then we cannot expect single answers. If the world is complex and messy, then at least some of the time we're going to have to give up on simplicities."[6] Animals—wandering, changing, dying—were part of the stuff of life that needed to be contained in a will, and the methods used to do this might begin to tell us much about testators and their relationships with their living possessions. Describing what is complex—trying to count chickens—is difficult, but my argument in this chapter is that early modern men and women often did a pretty good job of it by finding ways to accommodate the conceptual (not to say actual) messiness of living with and bequeathing animals. Unlike the social science studies in John Law's analysis, many of the testators in the early seventeenth century seem to have found ways to include the ephemeral, to maintain a kind of control over what was beyond their reach.

Crops have two statuses in wills and inventories: those in the ground are one thing, and those that have been harvested are another. So, included "in the Barne" in the 1626 inventory of the goods of Roger Newton, a great-grandfather from Good Easter who listed himself as a blacksmith in his will, is "wheat Barley and pease and some hay and strawe" as well as "two Cowes and Two hogges" Also listed, but separately from the other crops, was "the wheat growing vpon the ground."[7] The two groups of crops fit two different categories that are frequently found in wills. Those in the barn were transportable at the time the inventory was drawn up ("markettable" is a term used in the 1626 will of Henry Abell of South Hanningfield[8]), whereas those still growing in the field were not. This fact is reflected in the legal terms used

---

6. John Law, *After Method: Mess in Social Science Research* (New York and Abingdon: Routledge, 2004), p. 2.

7. ERO, D/ApgW 1/14 (1626), made probate 1,778 days after writing.

8. He included a bequest of "good clean sweet and markettable wheat." ERO, D/ABW 47/208 (1626), made probate 36 days after writing.

to represent such differences: "movable" and "immovable." Thus, we have, in various wills, for example: "all singuler my moveable goods," "goods & moveables," "moueables and goods," "moveable goods," and simply "moveables."[9] We also see "goods moveable and unmoveable" and "moveables and ymmoveables."[10] Movable goods are more frequently referred to than immovable ones in what I call the collecting phrase at the end of a will in the sample dataset of eighty-nine wills (twenty-three times compared to eight; 26 percent and 9 percent) simply, I assume, because while the term "movable" includes everything that can, literally, be moved, immovable goods are the land and the property that exists only in situ and thus are more valuable and much more likely to have already been specifically bequeathed in the preceding bequests in the will.

But "movable" and "immovable" are not the only terms used to encompass the residue of possessions that an individual did not specify in his or her will, and the first part of this chapter traces the variety of terms used in wills that might—and might not—include animals. At stake here is not simply whether there were more animals than are listed in the 10 percent of wills that specify them, but how the legal language that was used in a will might allow us to see the ways testators of early modern Essex managed to negotiate the complex path between preparing for the future and knowing that that future was unknowable. The terminology used, I will suggest, did not—to use Law's terms—attempt to tidy a mess by producing a hygienic clarity; instead, it tried to include the messiness; and its ambiguity was absolutely central to its true value.

Thus, in addition to "movable" and "immovable," we also find variations in the translation of the Latin *bona et catalla*. So "all the rest of my goods and chattels" or a variation that used the terms "goods" and "chattels" together appears thirty-four times in the sample dataset (38 percent of the time). Examples include "goods and chattells moveables and vnmoveables,"[11] "the rest of my goods & chattels,"[12] "all my moveable goodes and Chattles whatsoever,"[13] and "all the rest of my goods, plate, housholdstuffe & Chattells

---

9. ERO, D/ACW 11/55 (1629), made probate 77 days after writing; ERO, D/AEW 16/300 (1620), made probate 51 days after writing; ERO, D/AEW 17/159 (1623), made probate 396 days after speaking; ERO, D/AEW 17/217 (1624), made probate 573 days after writing; ERO, D/AEW 19/270 (1634), made probate 126 days after writing.

10. ERO, D/AMW 1/238 (1634), made probate 34 days after writing; ERO, D/ABW 47/19 (1626), made probate 1,162 days after writing.

11. ERO, D/ABW 44/139 (1621), made probate 122 days after writing.

12. ERO, D/ABW 44/325 (1623), made probate 85 days after writing.

13. ERO, D/ABW 49/55 (1629), made probate 23 days after writing.

whatsoever."[14] A similar phrase, variously spelled, variously ordered, but including the word "cattle" occurs eighteen times, or 20 percent of the time. Examples include "my goods and moveables what soeur Cattle chattles readie money bonds bills housholdstuffe & plate";[15] "househouldstuffe; goods, Cattell and moveables";[16] and "my goodes cattels & all other thinges."[17] On one level, these phrases are utterly conventional—the frequent usage, particularly, of "goods and chattels," signals nothing more than what is left over: it performs the function of gathering up. But a closer look at these collecting phrases reveals a more complex problem. Dictionary definitions are often crude ways of viewing issues of terminology, but here the definitions underline the problem. What follows is teetering on the absurd—language fails to tidy, terminology does not clarify, meanings slide in and out of focus—but comprehending this slipperiness is central to understanding what early modern testators were doing when they wrote their wills in the ways they wrote them.

According to the *OED*, the Anglo Norman legal term *"catel"* was "superseded at an early period by the Parisian *chattel"*and came to mean "chattel," a possession, while, from the sixteenth century, almost the only use of "cattle" (in its various spellings—cattel, cattal, catel, catle, etc.) referred to livestock.[18] The separation thus seems clear: chattels are non-animal movable goods and cattle are animals, which, if they're movables, are also self-moving. This clarity seems evident in the 1625 will of Thomas Porter of Tendring, who directed his brother-in-law John Pilgrim to sell "eight fatted Cattell & two bulls."[19] The easy clarity of "cattle" in Porter's will is not the norm, however, and there is a blurring in the use of the term that is particularly significant when dealing with wills, for although after 1500 "cattle" came to almost always refer to animals, the *OED* notes that the "technical phrase 'goods and catells (cattals)' which survived till the 17th cent." is the exception. In this usage, "cattells (cattals)" could also mean possessions.[20] Thus, what looks like a clear separation of cattle from chattel may not actually be so distinct. Likewise, the *OED*'s definition 3 of the collective sense of "chattel" works to undercut any difference between animal and inanimate thing that the separation of *"chatel"*

---

14. ERO, D/ABW 49/183 (1629), made probate 167 days after writing.

15. ERO, D/ABW 44/14 (1622), made probate 36 days after writing.

16. ERO, D/ABW 46/84 (1635), made probate 108 days after speaking.

17. ERO, D/ABW 48/128 (1627), made probate 294 days after writing.

18. *Oxford English Dictionary*, s.v. "chattle"; *Oxford English Dictionary*, s.v. "cattle."

19. ERO, D/ABW 47/195 (1625), made probate 90 days after writing.

20. *Oxford English Dictionary*, s.v. "cattle."

from "cattle" that took place around 1500 might seem to have created. Taking evidence from a usage dated 1627—that is, from right in the middle of the period of the large ERO dataset—the *OED* proposes that chattel could also at that time be used to mean "livestock."[21]

The nonstandardized nature of early modern spelling underlines the problem and consequently the difficulty of separating chattel from cattle in this period. In the 1632 will of Edward Halles of Romford, for example, is written: "I make my wife full executor and giue her all my goods and chatells during her natural life and After her decease that goods and catells which she doth leaue to be equally deuidd betweene my two sonns."[22] Here we should perhaps see the shift from "chatells" to "catells" as indicating the changeable spelling of the period, a changeable spelling that also reflects the lack of fixity of the terminology—or perhaps reinforces it. Either way, the difficulty of differentiating cattle from chattel is made more acute because of it.

But, that is not all: while "goods and chattels," a pairing that appears in 38 percent of the sample dataset, might logically seem to support the 1627 usage of "chattels" as livestock (there are things—goods—on the one hand, and animals—chattels—on the other), the *OED* once again offers little help. Definition 9a for the noun "good" reads: "Personal property, possessions; *esp.* movable property. Cf. CHATTEL *n.*4," revealing that "good" and "chattel" may not be so distinct after all. And definition 9c for "good" confuses things still further, noting that in northern England in the seventeenth century the term "goods" could also refer to livestock.[23] We are looking at documents from Essex, which is in the southeast of England, however, so perhaps here "good" need not have that northern meaning, and thus, following *OED* definitions as appropriate for Essex usage, when the nuncupative will of William Cottise, a man who was born and lived in Heydon, opened with the words "I giue my cow and all my other goods," we should, perhaps, assume that his meaning was that the cow was one kind of thing and the goods another, not that a cow was one of the goods, which may all be livestock.[24]

But even if we can set aside the ambiguity of this northern usage of "goods" in Essex, we need to remember that the term "livestock" was not

---

21. *Oxford English Dictionary*, s.v. "chattel."
22. ERO, D/AEW 19/137 (1632), made probate 23 days after writing.
23. *Oxford English Dictionary*, s.v. "good."
24. ERO, D/ABW 52/35 (1633), made probate 17 days after speaking. The record of William Cottise's baptism on 1 February 1571 is in the Parish Register of Holy Trinity, Heydon, ERO, D/P 135/1/2 (1558–1755).

used anywhere in England until after the period of the large ERO dataset: the *OED*'s first recorded use is 1687 (although "liveinge Stock" was recorded in 1660).[25] But if "livestock" is not a term found in the large ERO dataset, "stock" on its own is, and its usage reveals, perhaps, why the later addition of "live" was a necessary clarification. Stock could be used to refer to just animals: thus Christopher Bufford of Ingrave specified "Tenn of my Cowes and twenty Ewes to be indifferently taken out of my stock";[26] Thomas Plumer of Layer de la Haye bequeathed to his grandson "one Cowe bullocke of two yeares owlde to raise him a stocke";[27] John Wharton of Mountnessing included "all my stock of Cattell and Corne vpon the grounds and in the Barnes";[28] and John Hawkins of Rayleigh's nuncupative statement read "he did give all his stock of cattle ^corne^ and goods whatsoeuer equally betweene his said wife and his said sonne."[29] If we take "cattell" as it is used in John Gyver of Little Canfield's will to mean non-animal things then a similar clarity in the meaning of "stock" can be traced: "all my stock & corne cattell Implements of husbandrie & all other my goods about the said ffarme & windmill therevnto belonging," but of course such a use of "cattle" is not definite.[30] Another ambiguous use of "cattle" that apparently means animals at one point and goods at another is found in Anthony Sherlocke's 1625 will that also, however, includes another (possibly) clear use of "stock": "All my cattell corne & stocke of cattell whatsoeur."[31]

But "stock" was also used to refer to things besides animals in this period, and in his 1632 will, John Sowtho of East Mersea used the term in a way that rather undermines the difference between animals and crops that seems to exist in other documents. It included the direction "all such Stocke of Corne Cattell and other moueables shall be sold after Mychaellmas next and the monie thereof made shall be layd out in the purchas of land."[32] John Wharton's phrase "all my stock of Cattell and Corne vpon the grounds and in the

---

25. *Oxford English Dictionary*, s.v. "livestock."

26. ERO, D/ABW 43/178 (1620), made probate 1,465 days after writing.

27. ERO, D/ACW 10/267 (1627), made probate 98 days after writing.

28. ERO, D/AEW 18/249 (1628), made probate 195 days after writing.

29. ERO, D/AEW 18/101 (1627), made probate 40 days after speaking. "Stock" was also a synonym for "swarm" when applied to bees. In his will, Thomas Ive of Harlow bequeathed his grandson Thomas, among other things, "a stock of beese." ERO, D/AMW 1/34 (1623), made probate 850 days after writing. I return to bees in chapter 4.

30. ERO, D/AMW 1/213 (1633), made probate 102 days after writing.

31. ERO, D/ABW 46/137 (1625), made probate 21 days after writing. Again, the nonstandardized nature of early modern spelling could be playing a role here.

32. ERO, D/ACW 11/149 (1632), made probate 107 days after writing.

Barnes" seemed to allow the meaning "my stock of cattle and my corn" (i.e. that the term "stock" applied only to the "cattle"), but Sowtho's will reminds us that the usage may be more ambiguous than that: that "stock" may be referring only to corn or to both cattle and corn—"my stock of cattle and my stock of corn." Indeed, the focus on the location of stuff included in the term "stock" in some wills seems to underline this: Thomas Petchey used the phrase "all my stocke wthout doors," listing "All corne in the barne or grownd all Cattell, as Horse, Cowes, Bullockes, Coltes, sheepe, Hoggs or whatsoeur else is wthout."[33] Thomas Pellett also used this phrase in his will of 1623: "all my stock wtout doors of Cattell and Corne vpon my farme."[34] An opposition seems possible here, not between animals and non-animal stuff, but between organic things (animals, crops) and manufactured ones (the household items that might be kept inside). But even as that distinction might look plausible, it should come as no surprise in this realm of sliding meaning to find that in his nuncupative will of 1629 Simon Woods, a husbandman from Birdbrook, referred to "all his household stuffe and stock as well wthin dores as wthout."[35] No term, it seems, was fixed in its use.

Without the term "livestock" to distinguish animals from other possessions (the possessions that are movable but not self-moving), the animals of the testators from early modern Essex, when not individualized or specified, were included in general categories. The cows, pigs, horses, sheep, hens, geese, and bees the testators owned were simply among the belongings to which a monetary value could be given and that could be bequeathed. The best equivalent of "livestock" in the large ERO dataset is, perhaps, the phrase "quick Cattell" that can be first found in Statute 27.HenVIII.CapVII, "For the abuses in the Forests of Wales." This law from 1535—almost 100 years before the wills in the dataset—referred to "any beasts or quick cattle" who had escaped from the forests.[36] The term clearly did not enter into common usage in Essex over the next 100 years, as it was only used twice in the wills in the large ERO dataset, once by Thomas Pole a yeoman from Little Clacton who, in 1626 listed the items he was bequeathing to his wife Elizabeth. These included furniture, linen, and also animals:

> one Mare Colt with a bald fface, now bigg with Colt. And Three Eawes
> with three young lambes by their sides Indifferently taken out ffrom the

---

33. ERO, D/ABW 49/205 (1629), made probate 92 days after writing.
34. ERO, D/ACW 9/163 (1623), made probate 28 days after writing.
35. ERO, D/ABW 49/187 (1629), made probate 39 days after speaking.
36. William David Evans, *A Collection of Statutes Connected with the General Administration of the Law* (London: Blenkarn, Lumley and Bond, 1836), 4:20.

rest of my sheepe and lambes. And one old Tirkie henn and halfe the young Tirkies that she shall bring fforth, and halfe the younge ffowles, or Pullery of eyther kynde that they shall bring fforth. And my will & mynd is that all & every gifte of quick Cattell before given & bequeathed vnto her shalbe pastured meated and kept vpon the ffearme wheare I now dwell vntill the ffeast of St Michael Thearkangel next coming after my decease.[37]

The feast of St Michael the Archangel, also known as Michaelmas, was on 29 September and was used throughout the period to mark the end of the harvest and thus the end of the agricultural year. In the only other instance when the phrase "quick cattle" was used in a will in the large ERO dataset, William Drake of Great Waltham summed up his wishes with this statement: "And as for all the reste of my Goods and quicke cattell ^& Chattells^ I Giue them vnto Ellinor my welbeloved wife and Wm my sonn."[38]

In their use of this phrase, these two wills, even as they made clear that "cattle" can refer to animals, also raised the specter that there might be such a category as "dead cattle."[39] It is unclear whether such "dead cattle" might refer to goods in general or just to things like the (literally) dead animals that are the source of the "three bakon flitchis" that William Smieth of Great Sampford bequeathed to his wife Lediay.[40] But perhaps the fact that "dead cattle" does not appear at all in the large ERO dataset means we should discard that idea and accept that "quick cattle" is an early seventeenth-century equivalent of "live-stock" and that even though it was rarely used in wills was part of the language of the residents of Essex during the period of this study. The phrase worked, when it appeared, in opposition, not to "dead cattle" but to "cattle" as that lat-ter term had the potential to mean things in general. As if to confuse matters

---

37. ERO, D/ABW 47/15 (1626), made probate 113 days after writing.

38. ERO, D/AEW 19/267 (1634), made probate 279 days after writing. A search of the corpus of early English books on http://corpus.byu.edu/eebo revealed that "quick cattle" (search: qui*cat*) was used three times before 1660: twice in a proclamation "Of Regrators, forestallers, and Ingrocers. An. v. Ed. vi. Cap. xiiii," which noted that anyone found "Bying of quicke cattell, & sellyng of the same within .v. wekes, shall lose the double value" (dating from 1561 and 1562), and once in Leonard Mascall's *First Book of Cattell* (1587), in which he warned that pigs fed on the inwards of animals might develop a taste for them and be led to chase down "quicke cattell" to eat them. Early English Books Online, http://corpus.byu.edu/eebo/, accessed 27 July 2016. I use the term here both as a useful alternative to "animals" and "livestock" and because it underlines the complex position of agricultural animals in wills.

39. "Deadstock" did not enter usage until the 1830s. *Oxford English Dictionary*, s.v. "deadstock."

40. ERO D/ABW 44/315 (1623), made probate 37 days after writing. "Two small flytches of bakon" worth 5s also appeared in the inventory of the goods of Nicholas Garret of Boxted. ERO, D/ACW 9/93 (1622), made probate within 21 days of speaking.

yet again, however, the 1627 will of William Peacoke, a yeoman from Steeple Bumpstead, includes a bequest to his wife of "all my beasts & other my quick goods my haye & corne."[41] Lack of a stable punctuation system in writing of the period makes it difficult to see what "quick goods" refers to here: did it mean the other animals ("all my beasts and other quick goods, and hay and corn"), or did it refer also to the crops ("all my beasts and other quick goods—such as hay and corn")? Whichever usage seems likely—I favor the first—the phrase returns us not to the idea of dead goods (none of these are referred to in the large ERO dataset either) but to the sense that goods too can be quick, can be animals, even in the southeast of England.[42] By implication, this usage reveals, yet again, that a clear differentiation in terminology that might separate animals from insensible objects cannot be found in the collecting phrases of these wills.

One other term used in wills is worth pausing over here. It works in a slightly different way from "goods," "chattels," "cattle," and "stock." Unlike those terms, the word "appurtenances" is not used in the final collecting phrase that brings together the residue of stuff not already bequeathed in a will; rather it is used in relation to property. And "appurtenances" is used in almost 48 percent of wills in the large ERO dataset that include legacies of land or houses. A typical usage is present in Nicholas Merrell's bequest to his grandson Nicholas of "all my Customary & Coppihold messuage tenement & sixteen acres of land more or less with thappurtenances" in his 1624 will,[43] and in the yeoman Richard Radley of Great Hallingbury's bequest of eight acres of land to his son "wth all and singuler the appurtenances thereunto belonginge" in 1627.[44] In legal terminology, an "appurtenance" is, as the *OED* puts it, "a thing that belongs to another, a 'belonging'; a minor property, right, or privilege, belonging to another more important, and passing in possession with it; an appendage."[45] As Sara Birtles has noted, "An appurtenant grazing right may relate to any sort of animal, limited in number to those *levant* and *couchant*"; that is, to domestic rather than wild beasts.[46]

---

41. ERO, D/AMW 1/74 (1627), made probate 533 days after writing. A search of the corpus of early English books revealed four uses of "quicke goods" (search qui* good*) in the period 1590 to 1635. No other spelling variants of the phrase was found. Early English Books Online, http://corpus.byu.edu/eebo/, accessed 27 July 2016.

42. No parish register for this period exists for Steeple Bumpstead, so it is not possible to confirm that William Peacocke was born in the village and was not an incomer from the north.

43. ERO, D/ABW 45/118 (1624), made probate 37 days after writing.

44. ERO, D/AMW 1/226 (1627), made probate 411 days after writing.

45. *Oxford English Dictionary*, s.v. "appurtenance."

46. Sara Birtles, "Common Land, Poor Relief and Enclosure: The Use of Manorial Resources in Fulfilling Parish Obligations 1601–1834," *Past and Present* 165 (November 1999): 83. I return to the idea of "levant and couchant" in chapter 3.

Thus, if the almost 48 percent of testators in the large ERO dataset who bequeathed property were using the term correctly, the appurtenances they refer to might be the grazing rights that were attached to the land they were bequeathing, and this could mean two things. First, that more than 48 percent of testators who bequeathed property also possessed appurtenances because, as appurtenances are not separable from the land they are appended to, they could pass automatically with it and did not actually need to be referred to in order to be bequeathed. This would mean, in turn, that the figure of 48 percent might underestimate the real number of property owners with appurtenant rights. And second—and consequently—it could mean that a far greater number of testators than has previously been evident in the large ERO dataset might own animals, because such possession would be inevitable, one would assume, if grazing rights were held.[47]

But this, of course, assumes that "appurtenances" was being used in its legally correct meaning in all the wills, and that is not something that should be taken for granted: and, indeed, the wills themselves, yet again, offer evidence of a wider usage. There are occasions, for example, when "appurtenances" is applied to the relationship of bedding to a bed and thus means generally "the stuff on the bed." John Basie of Great Birch in northeast Essex, for example, left his son John "my trundlebed standing aloft in the chamber wth the greene Rugge the flockbed and the rest of the appurtenances" and his other son Francis "my best Joyned Bedstead wth the appurtenances";[48] Nicholas Churchman of Wendens Ambo, in the northwest of the county, bequeathed Mary, his wife, "one bed and bedding in the Chamber wth all thappurtenc";[49] Steven Camber, a mariner from Hadleigh on the south coast, bequeathed "a featherbed & the appurtenaunces to it" to his daughter Mary;[50] Elizabeth Crispe of nearby Foulness—an island off the east coast—left "one bedsteddle & bedde with the appurtenances" to be equally divided between her son and daughter;[51] and James Stanes, a bricklayer from Boreham, in the middle of the county, gave each of his three sons "one fflockbedd wth the appurtenances therto belonginge."[52] The geographical and temporal spread of this usage of "appurtenances" makes clear that this was not

---

47. The exception here might be on land on which animals were leased by the landowner/tenant. On this see Elizabeth Griffiths and Mark Overton, *Farming to Halves: The Hidden History of Sharefarming in England from Medieval to Modern Times* (Basingstoke: Palgrave Macmillan, 2009).

48. ERO, D/ABW 49/100 (1628), made probate within 135 days of writing.

49. ERO, D/ABW 51/56 (1632), made probate 23 days after writing.

50. ERO, D/AEW 17/144 (1623), made probate 67 days after writing.

51. ERO, D/ABW 51/163 (1633), made probate 178 days after writing.

52. ERO, D/AEW 17/74 (1622), made probate 50 days after writing.

a simple case of one scribe repeating an idiosyncratic phrase across a number of the wills he had been asked to write.[53] It could be, of course, that these five testators or their scribes misused what is an otherwise correctly used term in their wills, but I think—especially given the wider lack of terminological clarity in what I have called the collecting phrases—we might consider that their usage signals something else: that it is possible that when "appurtenances" was used in relation to land what it refers to is simply everything on, and associated with, that land, and did not specifically refer to any appended legal right. It might, therefore, mean timber, for example, or the animals that are likely to be grazing there. Thus, it is possible that a cow might be listed as being in the barn in an inventory but might be classed as an "appurtenance"—that is, as being appurtenant to the land—in a will.

This is speculation, but what this speculation allows us to see is how potentially easy it was to not specifically mention animals in wills while including them in bequests (as cattle, stock, appurtenances, and so on). This in turn underlines how difficult it is to find these unspecified animals again from the distance of centuries. Another implication of my reading of "appurtenances" is that animals might have been owned by many more people than the 10 percent of testators in the large ERO dataset who included specific bequests of them. We can begin to think about another sizeable group of documents as also potentially (and silently) including animals—not only those that include the term "appurtenances," but also those wills that do not include the term "appurtenances" but do include property, because appurtenant rights would automatically be transferred with the land, so grazing rights might be present even if they are not specified. Eighty-four wills include both specific animal bequests and references to appurtenances, so we cannot simply add the more than 48 percent of wills that mention appurtenances to the number of wills that include animals to come to a total of almost 60 percent of wills in the large ERO dataset that perhaps include livestock,[54] but recognizing this possibility—and acknowledging the confusion of meaning in the terms

---

53. Another use of "appurtenances" was recorded in 1701, when Richard Gough noted that "if a man sell a dwelling-house with the appurtenances, the seate in the church passes by the word appurtenances." Gough, *The History of Myddle*, quoted in Matthew Johnson, *An Archaeology of Capitalism* (Oxford: Blackwell, 1996), p. 99.

54. It is interesting to note that a testator who used the term "appurtenances" was slightly less likely to include a specific animal bequest than one who did not use that term: 9 percent compared to 10.5 percent. This difference would seem to support the proposition that use of the term "appurtenances" implies the possession of animals and that because of this, specific bequests were often not regarded as necessary.

"cattle," "chattels," and "goods" that are used in so many documents—might mean that the wills begin to reflect more fully the rates of animal ownership that the inventories allow us to see. Thus, while only 10 percent of testators specify animals, perhaps we might recognize that the majority of others included them in ways that make them almost invisible.

This is not simply the problem of counting chickens in the yards, however, and here we enter conceptual territory. It is worth considering whether the inclusion of animals in the term "appurtenances" in a will signals a particular attitude toward them—an attitude that denies the animals individuality and regards them only as part of a system of agriculture—as moving crops or quick goods—or if we should view the language used in the wills as more of a reflection of the limitations of the conventions of the last testament. Indeed, it is worth considering whether those two—attitude and convention—can be separated. We need to recognize that language does more than simply reflect reality: as John Law writes, "Methods, their rules, and even more methods' practices, not only describe but also help to *produce* the reality that they understand."[55] The confusion inherent in the terminology used in the writing of wills might, then, seem to make more likely the objectification of animals. In turn, it might mean that the social groups that were more likely to write wills—the upper and middle orders—were therefore being invited more than their poorer neighbors were to think about their animals in general and collective rather than in individual and particular terms.

There is a bottom line, of course, that all testators shared: any animal included in a will—whether in a collecting phrase or as a specific bequest—was simply a thing of value. They were stuff that the testator owned and passed on as economic benefits to his or her legatees. In this formal legal context, the term "cattle" does not need to distinguish the quick from the dead and "stock" need not differentiate animal from vegetable. As such, the language seems to function simply to tidy—to put in one place what is scattered, both physically (in the yard or in the barn, without doors or within) and conceptually, and the collecting phrases used in wills (whether they used the expression "the rest of my goods and chattels" or the term "appurtenances") might be viewed simply as shorthand, as quick ways of saying something that might otherwise take a long time and use up a lot of (expensive) paper.[56]

---

55. Law, *After Method*, p. 5, italics in the original. Law is not denying that there is a reality outside of language.

56. It is notable how many wills are written to fit onto one page. In the sample dataset of eighty-nine wills, seventy (79 percent) are one page long; thirteen (15 percent) are two pages long; four (4.5 percent) are three pages long; and two (2 percent) are four pages long. The desire to remain on one

But assuming the language of the collecting phrase was simply an attempt at succinctness avoids attending to those phrases' lack of inherent clarity as they were actually used (rather than as they might be defined in legal works). It is possible that the collecting phrases, and their apparent objectification of animals might actually be doing something more interesting than allowing for brief legal documents.

Law writes a list of the "textures" of life that, he argues, are missed out by the desire for clear description in social science research:

> Pains and pleasures, hopes and horrors, intuitions and apprehensions, losses and redemptions, mundanities and visions, angels and demons, things that slip and slide, or appear and disappear, change shape or don't have much form at all, unpredictabilities.[57]

Such a list can be used to think about the writing of an early modern will, but I suggest that, in the act of putting down on paper or parchment one's final wishes, the messiness, the stuff of which life is truly made—the connections, engagements, characters, habits, moods, shifts, flux—is not excluded, but remains and, indeed, that its remaining is vital to the carrying out of those wishes. The collecting phrases are not simply objectifying clarifications, they are a way of making the mess of life present in a coherent way, and a look at the religious and social pressures that underpinned writing or dictating what should happen to one's property after one's death at this point in history reinforces this interpretation. In such a context, the collecting phrases can be understood to bear more fruit (to use that horticultural metaphor) than initially might seem apparent to a modern reader.

This book focuses on one historical moment of will writing, so the specificities of that moment are important, and the lengthy preamble to the 1625 will of John Wall, a yeoman of Great Hallingbury, offers a helpful introduction to a key conception of the will that was held in that period and that must influence our interpretation of these documents.

---

page can also be seen in documents in which the final few lines are squashed into the remaining space in ways that can make them difficult to read. See, for example, ERO, D/ABW 49/357 (1627), made probate 251 days after writing; and ERO, D/AMW 1/83 (1625), made probate 23 days after writing. Longer wills do not always contain more bequests: frequently they are written in a larger hand with more generous line spacing and contain little more than one- or two-page documents do. The wealth of the testator might thus be mirrored in the writing of their will on numerous sheets of paper.

57. Law, *After Method*, p. 2.

Being in good health of boddy, & of sound & pfect mynde and memorie laud & praise be given to god calling to mynd the fraile & fickle estate all men have here in this transitory world not knowing howe soone wee may be called hence and being desirous to settle and dispose of such worldlie estate as it hath pleased god to bestowe vppon me wherewth I am desirous not to be troubled or encombred when it shall please god to visite me wth sicknes doe therefore in the tyme of this my health make & ordeine my laste will and testamt in manner & forme following.[58]

Taking seriously the declaration in the burial service in *The Boke of common prayer* that "in the middest of life we be in death," individual Christians were urged to prepare for their end, and writing a will, as Wall knew, was one aspect of that preparation.[59] Indeed, Ralph Houlbrooke notes that the Church taught that writing a will "was as much a duty as a right, an expected part of the Christian's preparation for death, and a powerful means of helping the soul."[60] To leave the declaring of one's last will and testament until one was on one's deathbed was to be distracted by earthly matters when one should be focusing on the eternal. As Robert Maynard of Little Bromley put it in 1622, "considering the fraile estate of man, and that euerie sickness is a warning to death, [I] doe make ~~this~~ and ordaine this my last will and testament."[61] A last will and testament served a valuable function, then. In terms of its spiritual work, it was a preparation for death—a kind of legal memento mori.

The anonymous 1612 text *A Pithie and Short Treatise by Way of Dialogue whereby a godly Christian is directed how to make his last Will and testament* quotes Isaiah 38:1 on its title page ("Put thine house in order, for thou must die") thus setting this work firmly within the context of the Protestant views cited above. But this text moves things in a rather different direction and asks readers to recognize how dangerous it would be to forget this world in preparation for the next. The dialogue that makes up the main text of the *Treatise* has "Minister" say to "Christian": "you haue done full euill, in not making your will in the time of your health, seeing your old age so to approach, that by

---

58. ERO D/AMW 1/71 (1625), made probate 179 days after writing.

59. "The order for the Buriall of the dead," in *The Boke of common prayer and administration of the Sacramentes and other rites and Ceremonies in the Church of Englande* (London: Edwardi Whytchurche, 1552), Piiiv.

60. Houlbrooke, *Death, Religion and the Family*, p. 81. See also Richard Wunderli and Gerald Broce, "The Final Moment Before Death in Early Modern England," *The Sixteenth Century Journal* 20, no. 2 (1989): 259–75. They suggest that the belief that one's final moment might impact one's eternal state was popular but "ran counter to all theologies of death" (p. 261). I suggest that the emphasis on writing a will before illness struck was one in which the theological and the popular might coincide.

61. ERO, D/ACW 9/104 (1622), made probate 29 days after writing.

the course of nature your daies could not be long." The reason Minister gives for the importance of this task is not linked to the well-being of Christian's immortal soul, though; it is about very mundane things. Minister reminds Christian, "if God should haue taken you away sodainely . . . then should you thereby haue bene perhaps the Author of bitter strife and contention amongst your children; which is one of the things so abhominable & hateful in Gods sight, namely to raise contentions among brethren."[62]

Thus, a will performed temporal work for the testator and also supported his or her efforts to secure a stable future for their family when they would no longer be there to help in person. Indeed, in his 1620 will, Samuel Parish a mariner from East Donyland voiced both desires—to deal with the material realm in order to focus on the spiritual and to avoid contention in the world that he left behind:

> considering the frailtie & mortaletie wherevnto man is subiect & the vncertayne continuance of lief and state being resolved against all suddaine alteracons that may happen in this earthlie condcon to prpare my mynde to a ~~sed~~ setled course of quietnes & meditacons of heavenly comforts and ioyes and To determine of the dispiosicon of those lands and goods wherewth the Lord of his mercy hath blessed me so as peace to mine owne sowle and quietnes amongest my children may after my death be firmely kept & maynetayned doe make & ordaine my testamt contayning my last will in manner and forme followinge.[63]

Here was recognition that writing a will was an act of both spiritual and social significance.[64]

Whatever the underpinning wish of testators, John Law's reading of the desire for clarity allows us to see that the kind of precision apparently offered by a will—its gathering up of the scattered, its bundling together of the disparate, its representing as complete what was frequently changing—might silence some of the indispensable and inevitable messiness of day-to-day life.

---

62. *A Pithie and Short Treatise by Way of Dialogue whereby a godly Christian is directed how to make his last Will and testament* (London: William Iones, 1612), p. 3.

63. ERO, D/ACW 8/261 (1620), made probate 56 days after writing. Lucinda McCray Beier notes that this "regard both for the next world and the world left behind" was part of an "ideal seventeenth-century death." Beier, "The Good Death in Seventeenth-Century England," in *Death, Ritual, and Bereavement*, edited by Ralph Houlbrooke (London: Routledge, 1989), p. 52.

64. It might be suggested that it was a scribe who included this preamble, but the fifth and final page of Parish's will challenges this suggestion. Here Parish offered advice to his children to "keepe faith . . . love & concord amongest themselves . . . keepe themselves vnspotted." Giving such advice in a will was unusual and might signal that it—and so the sentiment in the will's preamble—came from the testator himself. ERO, D/ACW 8/261.

What we must also acknowledge when it comes to wills, however, is that that tidying was, in a sense, exactly their function. Wills intended to give clarity where dissolution was possible. Indeed, they might be said to cover over what was actually their raison d'être—the temporariness of human possession—with the impression of the existence of perpetuity and continuance. Entailing property was an attempt to ensure uninterrupted ownership, to maintain the status quo: the King is dead, long live the King. In making possession seem "absolute," to use a legal term, wills disguised from view the fact that if in the "middest of life we are in death," then ownership is always already "qualified," tenure transient, proprietorship impermanent.[65] As such, bequeathing one's property appears to hold firm what are always shifting sands.

In this way we can see that wills by their paradoxical nature (their attempt to make absolute what is inherently qualified) could never encompass the apprehension, loss, disappearance, change, and unpredictability that attended death, whatever wording a testator chose. And this is not simply because language is slippery. It is more because a will attempts to construct as completed what could not be so at the time of writing or speaking. Only death could mark a true end, but "a true end" is only such for the testator: for those they leave behind, the world and its messiness goes on. This is why inventories rather than wills seem to offer fuller pictures of actual possessions. They mark a stop to movement: here, today, this is what this person possessed.[66] "A henne and ten chickens . . . 5 chickens . . . 2 more." But wills, of course, contain more than inventories, and if they do not include a complete list of property, what they do offer are glimpses of a world that is otherwise very difficult to trace: a world of personal relationships—of bigger boys and little boys; of kinship and neighborhood networks in which one man can take the inventory of another

65. On "absolute" and "qualified" possession, see Michael Dalton, *The Covntrey Ivstice, Conteyning the practise of Ivstices of the Peace out of their Sessions* (London: Society of Stationers, 1618), p. 234.

66. But even inventories contain ambiguities and omissions. Goods might be taken before the list was made, objects might be misrecognized, valuations might be overgenerous. Carole Shammas also notes that "not all inventories have accompanying papers and not all of the papers contain the same kinds of information" and that decedents for whom inventories were created "are not necessarily representative of the living population." Shammas, "The Determinants of Personal Wealth in Seventeenth-Century England and America," *Journal of Economic History* 37, no. 3 (1977): 676. J. A. Yelling notes that inventories "do not say much about the use for which animals were kept; whether, for instance, oxen were for draught, meat or both." Yelling, "Probate Inventories and the Geographies of Livestock Farming: A Study of East Worcestershire, 1540–1750," *Transactions of the Institute of British Geographers* 51 (November 1970): 119. For a crucial challenge to the apparent objectivity of inventories, see Lena Cowen Orlin, "Fictions of the Early Modern English Probate Inventory," in *The Culture of Capital: Property, Cities, and Knowledge in Early Modern England*, edited by Henry S. Turner (New York and London: Routledge, 2002), pp. 51–83.

and then, three years later, witness his widow's deathbed testament. For this reason wills are better for qualitative than quantitative assessment, even if the quantitative can reveal some interesting trends.[67]

To return to the focus on animals, including them presents particular difficulties for the testators that go beyond what terminology to use. Individual Christians were urged to write a will in advance of illness in order to avoid being "encombred when it shall please god to visit me," as John Wall put it, and the will had, by logic of course, to be written before death. It was thus inevitable that however short the time between writing a will and enacting it, its contents would have changed: time would have passed, possessions would have altered, legatees might have died. Clarity falls apart as soon as it is inscribed and in such a context it becomes possible that the collecting phrases that might well be where animals are most frequently to be found in the wills are able to offer a more accurate picture of a life and its possessions because of all their ambiguity, because they are not fixed. In avoiding clarity, what might look like a failure might actually help the testator achieve what they intend after their death. And the lack of clarity might also evidence not the objectification of animals but their particular natures and how important bequeathing them was.

This might sound paradoxical, but I suggest that it is in the moments when the unknowability of possession is most fully present that wills recognize ambiguity as the better means of expression. To quote Law again, "If much of reality is ephemeral and elusive, then we cannot expect single answers. If the world is complex and messy, then at least some of the time we're going to have to give up on simplicities." In the context of this discussion, "simplicity" does not mean the use of the term "appurtenances," it might in fact mean the inclusion of a named or clearly described cow—a moment in a will with only one meaning: that heifer, that one, there: Nan. Such "simplicities" are meaningful and I return to them in chapter 3, but here I want to look at the vagueness of wills—at, to follow Law's terminology, their moments of complexity. Remember, only just over 10 percent of wills in the large ERO dataset include specific animal bequests: that means that just under 90 percent of those wills include either no animals at all or only the imprecision of collecting phrases. Here, finally in this chapter, I want to suggest that those collecting phrases do more than save expensive paper. I want to propose that they reflect something

---

67. Samuel Cohn Jr. makes a similar argument about the value of wills over inventories in his study of Renaissance Italy. Cohn, "Renaissance Attachment to Things: Material Culture in Last Wills and Testaments," *Economic History Review* 65, no. 3 (2012): 984–1004.

significant about early modern life that might be easy for a modern reader to miss: that the lack of precision of the collecting phrase might be where apprehension, loss, disappearance, change, unpredictability—those "textures" so important to everyday life—might best be traced, and it is the animals that lay this bare.

The agricultural year dictated the lives that are represented in many of the wills in the large ERO dataset. Whether that was in mentions of Michaelmas, the mark of the end of the harvest and of the agricultural year, or in reference to what was available to be bequeathed at different points in the calendar, testators were brought face to face with the changeableness of life in a way that many of us now are more distanced from. This is not just because "medieval and early modern Europeans . . . typically experienced the deaths of family and community members in far greater numbers than their modern counterparts," although that would certainly have had an impact.[68] It was also because time moved, and the world around one altered in ways one could not control or stop: that was the nature of existence in rural England. For example, in her 1624 will, Mary Butler, a widow, included an acknowledgment of the uncontrollable nature of life and death and the impact of that upon bequeathing:

> yf yt shall please god to take me out of this lyfe after my hopp-groundes in Margaretting shalbe digged or poled and before my hoppes thereof comeing shalbe gathered or picked . . . then my sonne James to whom the said lands are to come after my decease shall suffer my executrix [his sister Elizabeth] . . . to retayne and hold the said hopp grounds vntill Michaelmas next following after my decease and to gather and take the said hopps to her owne vse wthout the let or hinderance of my said sonne James.[69]

Death is movable, harvests are not, and the conditional phrase—"yf it shall please god"—is necessary, even in a declarative document.

Another example, this time emphasizing not the testator's own death but that of a legatee, can be found in the 1623 will of the husbandman John Parker of Farnham. A clear statement—"I gyve and bequeath vnto Mary my wyfe All my lands ^both free & coppie^ and my debts"—is followed by "But yf Mary my wyfe shall happen to die before the next harvest or before she hath

---

68. Rebecca F. McNamara and Una McIlvenna, "Medieval and Early Modern Emotional Responses to Death and Dying," *Parergon* 31, no. 2 (2014): 2.

69. ERO, D/AEW 17/186 (1624), made probate 865 days after writing.

satisfied and paide my debts Then my will and meaninge is that my sonne John Parker shall haue the cropp of Rye now growinge vpon a crofte of lande lyinge on the backside of my howse and tenn sheepe and lambes wch are pcell of my goods."[70] Parker's will is dated 28 March 1623, he was buried on 3 April, and his will was made probate on 5 May,[71] and even in the short period between these dates the rye would have continued to grow and the lambs to increase in size. Time would have passed before the will was proved and, just as legatees might die, so bequests did not cease to change during the process of executing a will.[72]

The unknowable nature of mortality—the fact that death might occur at any time—had particular implications for animal bequests, and this was recognized in wills too. Thus, John Thresher a yeoman of Childerditch, whose will I look at in more detail in chapter 3, bequeathed to his wife Phoebe "Ten milche Cowes to be Chosen by her self out of such kyne as I shall dye possessed of; or ffyfie pounde in money att her choyse."[73] "Such kyne as I shall dye possessed of" seems to be the opposite of clarity: it avoids specificity, but it also avoids designating a figure that time, poor harvest, illness, or age might impact in the period between the writing of the will and death. A variant of the phrase Thresher used appeared in the will of another yeoman, John Normanton of Cranham, who bequeathed "ffowre of my cowes, whereof I shall dye possessed" to his wife Johan. He became more definite immediately after this, however, bequeathing her, in addition to the cows, "sixe sheepe: one horse with his furnyture." The clarity here may have been easy: he may have had many more than six sheep and more than one horse. All we know about the other animals—the other cows he died possessed of and the possible sheep and horses—is the collecting phrase at the will's end: "All the rest of my goods & chattels vnbequeathed I give & bequeath vnto my said sonn John whome I make Appointe & ordeyne my sole executor of this my laste will & testament."[74]

---

70. ERO, D/ABW 44/285 (1623), made probate 38 days after writing. "Pcell" is an abbreviation of "parcel." The term appears to be used here as it is still used in the phrase "part and parcel"; definition 2b in the OED is "an integral or component part or member (of something). Used to emphasize inclusion in the whole, rather than partitive character; often without article." Parker's wording seems to follow this usage and to regard the sheep and lambs as part of the land.

71. For his burial date, see Parish Register, St Mary the Virgin, Farnham, ERO, D/P 290/1/1 (1559–1635).

72. The burial of Mary Parker, John's wife, took place on 16 December 1623, so she survived "the next harvest" by a few months. Parish Register, St Mary the Virgin, Farnham, ERO, D/P 290/1/1.

73. ERO, D/AEW 19/102 (1631), made probate 705 days after writing.

74. ERO, D/ABW 49/357 (1627), made probate 251 days after writing.

The changeable nature of the animals a testator might own emerges again in the will of Lettice Lynn, a wife who was in a position to entail property. In her 1622 will she left "all my lands and tenements whatsoever both free and coppie as they now lye and bee" to her husband William, but only until his death. Then the property was "to come to and descend vnto Richard Bastick sonne of John Bastick my kinsman" or, if William died before Richard reached twenty-one, it was to be placed at the "disposing and ordering of Sir Harbottell Grimeston of Bradfeild" until Richard reached his "full age." In addition to the entailed property, Lettice also bequeathed Richard Bastick "Eleven ~ milch beasts or cowes and thirtie and two sheepe" plus a silver bowl and salt, six silver spoons, a gold seal ring, her best featherbed and bolster, a pair of pillows, a pewter basin and ewer, eight cushions, some curtains, a table with two leaves, a cupboard with two drawers, and a brass pot. Lynn added: "All wch I will and ordeyne all as they now are wthout anie diminishing or changing the cowes and sheepe onlie excepted." Silverware, perhaps, could be put in a chest to guard it, and soft furnishings protected under covers; but animals cannot be stopped from "diminishing or changing." The will continues:

> And because the time of my said husbands death is vncertaine, and for that the pperties of the said cowes and sheepe may bee altered I doe further will and ordeyne that it shalbee at the choyce of the said Sir Harbottell or his Assignes whether the said Sir Harbottell or his Assignes will take the said Eleven cowes and thirtie and two Sheepe or the sum of fowrtie and eight pownds of lawfull Englishe money in consideracon and lew of the said cowes and shepe.[75]

Uncertainty and changing properties are the bane of any will, but they are also unavoidable. Indeed, it is the changing properties of a legatee rather than a legacy that is a focus for John Holmes, a cook from Tilbury-juxta-Clare. In his 1626 will he bequeathed two cows, one to each of his daughters, Elizabeth Harrington and Katherine Moore. But he also declared that his wife Ann should have "All the profit and benefit of . . . twoe Cowes during the term of her naturall life." The animals were, like land, entailed. Holmes's will continued: "yf the said Ann my wife (by reason of sicknes blindness or lameness fallinge vpon her) shall be hereafter enforced to sell awaye any of the Cowes aforesaid That then the said Elizabeth and Katherine shall loose her or their

75. ERO, D/ABW 44/173 (1622), made probate 153 days after writing. The tilde - ~ - is used to fill a space at the end of the line to confirm that no other word was intended and omitted.

partes [that is, lose the legacy] otherwise not."[76] It is likely that they did not "loose her or their partes," however, as Ann Holmes's burial is recorded in the parish register on 20 February 1627, less than six months after her husband's.[77]

Clerical encouragement to write wills in advance of one's final sickness had inevitable consequences for testators, then. How to second-guess providence? How to know who would be there to leave things to and what would survive to be bequeathed? The problem was not just how to count one's chickens after they have hatched, but how to count them before. In this vein, Barbara Clarke, a singlewoman from Buttsbury, wrote in detail about her wishes regarding one of her cows. Having bequeathed to her brother Andrew Clark's son "one Cow of colour black," she made a bequest to her niece:

> to Priscilla Dolor my sister James daughter one Cow my youngest browne cow with ye weynell [i.e., weaned animal] brought vppe of a calf to be de hers imediatly after my decease, but my will is that my brother Andrew shall have the vse possession and pfit of the sayd cow, for one wholle year giving to the father of the sayd Priscilla xiid for the sayd Cows hyre, and my sayd brother Andrew shall have the sayd Cow, if he shall at the end of the sayd year paye vnto William Dolor the father of the sayd Priscilla ^fower pounds of lawfull English money^ at the howse of John Norton of Stock clark if he then ^be^ lyvinge or els in the church yard of Stock alias Harvardstock & that if the sayd Cow shall miscary, or be dead at the end of the sayd year, then my sayd brother shall give fower poundes of lawfull money to the sayd father of the sayd Priscilla.[78]

"If he then ^be^ living"; "if the sayd Cow shall miscary": such were the vagaries of life. A testator could not prevent them; the best he or she could do was attempt to account for such possibilities.

The husbandman John Archer of Bradfield seemed even more reconciled to the presence of mortality—his own as well as that of others. Using Michaelmas as his key date, his will consisted of a preamble and the following:

> I give & bequeath unto Michaell Archer my sonne one Cowe called the brinded cowe to be deliu'ed to him or to his assignes att & vppon the nine & twentie daie of September next insueing the date hereof or

---

76. ERO D/AMW 3/196 (1626), made probate 258 days after writing.

77. Parish Register, St Margaret, Tilbury-juxta-Clare, ERO, D/P 164/1/1 (1560–1724).

78. ERO, D/ABW 47/31 (1626), made probate 36 days after writing.

ffortie shillings of lawfull mony of England, if the foresaide cowe doth chance to decaye in the meane tyme. Item I give & bequeath vnto Nicholas Archer my Sonne one Cowe called the little white faste [i.e., faced] brinded cowe to be deliu'ed vnto him or to his assignes att & vppon the nine & twentie daie of September next ensueinge the date hereof or ffortie shillings of lawfull mony of England, if the foresaid cowe doth chance to decaye. Item I give and bequeath vnto Rebecka Archer my daughter one Cowe called the little black Cowe to be deliu'ed vnto her or her assignes att & vppon the nine & twentie daie of September next ensueinge the date hereof, or ffortie shillings of lawfull money of England if the foresaide cowe doth chance to decaye in the meane tyme. Item my will & mynde is that if any of my children fortune to die att & before the said nine & twentie daie of September that then thee legacie to them bequeathed to be equeally divided betwin the survivors of them.

Everything else—"my goods and chattels vnbequeathed"—he left to his "well beloved wife" Ann, whom he made his executrix.[79] The struggle of a will writer, then, is evident: how to leave clear directions as to the dispersal of one's estate in advance of death and simultaneously attend to the real possibility that such clarity would make no sense by the time the will came to be read. John Archer got around this struggle by including it in his will and offering monetary alternatives to the specific animals he bequeathed; John Thresher's "Such kyne as I shall dye possessed of" is another method; "all my other goods & househoulde & cattell" is, perhaps, the most common one.[80]

In the will of Richard Freebody, a gentleman from Little Stambridge, another difficulty emerges, however, that highlights the issues this chapter has been focusing on: what it means to generalize about animals in wills. Some testators made bequests for prospective enactment: that is, they included gifts that should be put in action a number of years after their death. Thus Freebody bequeathed to "Ric Martin sonne of Ric Martin aforesaid one Ewe lambe to be deliu'ed vnto him foure yeares hence if he live soe longe."[81] The will is dated 20 January 1621, which makes it likely that lambs were not quite being born as it was signed. But lambing would have begun by the date of probate—14 February of the same year. In *Five Hundred points of good Husbandry,*

---

79. ERO, D/ABW 47/212 (1626), made probate 78 days after writing.
80. ERO, D/ACW 11/78 (1629), made probate 16 days after writing.
81. ERO D/ABW 43/71 (1621), made probate 25 days after writing. A copy of this will is at ERO D/AEW 17/4 (and is not included in the large ERO dataset because it is a copy). The existence of two versions in two different jurisdictions (ABW and AEW) tells us that Freebody had property in each.

Thomas Tusser advised the husbandman in January that "Yong broome or good pasture, thy ewes doe require / warme barth and in safety, their lambes doe desire."[82] In his nineteenth-century edition of Tusser, William Fordyce Mavor notes that "a barth is a small sheltered enclosure near the farm-house, where the ewes about yeaning time, are brought for safety, and more easy superintendence."[83] So January was a month to prepare for lambing. Included in his directions for activities to undertake in February, Tusser wrote:

> Keepe sheepe out of bryers,
> keepe beast out of miers.
> Keepe lambes from fox,
> else shepherds goe box.[84]

Freebody's bequest to young Richard Martin would work in terms of Tusser's farming calendar—four years from the date of the testament would be about the time the lambs were appearing. That much was sure. What was less certain was whether the son of Freebody's friend would be alive at that time. The generic bequest—the lack of specificity as to this "Ewe lambe" or that—was logical: the "Ewe lambe" bequeathed had not been born at the time of the will's writing. The legatee, on the other hand, had to be specified. Henry Swinburne wrote that an error in naming a legatee would mean that "the disposition is voide." He gave an illustration: "to giue Iohn at Stile an hundred pound, he saith, . . . giue I to Iohn at Noke an hundred pound. In this case, neither can Iohn at Stile, nor Iohn at Noke . . . obtaine the legacie. The reason is this: Iohn at Noke is excluded, because the testator neuer thought it. Iohn at Stile is excluded, because the testator never spoke it: for meaning without speaking is nothing, and speech without meaning is lesse."[85] It is logical that an unborn lamb cannot be specified whereas a living boy can. However, when the generalization is juxtaposed with the necessary rider about the particular legatee in the bequest to Richard Martin the younger—"if he live soe longe"—this makes the status of the animal seem more robust

---

82. Thomas Tusser, *Five Hundred points of good Husbandry* (London: I. O., 1638), p. 63.

83. Thomas Tusser, *Five Hundred Points of Good Husbandry*, edited by William Fordyce Mavor (1812; repr., Cambridge: Cambridge University Press, 2013), p. 92n3.

84. Tusser, *Five Hundred points* (1638), p. 67.

85. Henry Swinburne, *A Brief Treatise of Testaments and Last Willes* (London: Iohn Windet, 1590), p. 244v. Swinburne does not explain how an executor or executrix might know that the words spoken were actually misspoken. If the misspoken words were inscribed in the written document they would stand, because the only person who could correct them—the testator—would be unavailable for comment at the point when legacies were distributed.

than that of the human. A general category can persist, can transcend physical change and decay, while an individual might disappear forever. "If he live soe longe" sounds like a kind of forlorn hope.

As Freebody's bequest signals, general categories were useful. And the law recognized this in another way too. Specificity was actually dangerous when it came to things being bequeathed—especially if they had the tendency to "decaye." Swinburne writes:

> If the thing bequathed do perish or be destroied, the legacie is extinguished, and the legatarie destitute of remedie: For example; The testator doth bequeath vnto thee his best oxe, which oxe is afterwards killed: In this case the legacie is extinguished, insomuch that neither the skin, nor the flesh, nor the price, is due vnto thee.[86]

In opposition to this is "the legacie [which] is general, or consisteth in quantitie." A general bequest would be "when the testator dooth bequeath a horse, or an oxe (not this horse, or that oxe:) . . . this kinde of legacie cannot perish, though all the testators cattell do perish . . . and therefore the legatarie may recouer his legacie."[87] Here the recipient of the bequest is protected by generalization, by the refusal to specify one decaying being over another. However, Swinburne's discussion of the bequest that "consisteth in quantitie" reveals another side to specificity. A legacy that specifies a number of animals, Swinburne writes, "cannot perish, though all the testators cattell do perish."[88] This, again, protects the legatee from the ravages of dearth, flood, and illness among cattle, for example. But, by implication, a will that bequeaths a specific number of animals requires the executor or executrix to fulfil such a bequest, even if it means them having to buy new animals to do so. And fulfilling such a bequest could leave the executor or executrix—almost always a close family member[89]—in dire financial straits. To go back to John Normanton's bequest of six sheep to his wife: if he had died with only two sheep, his executor, his son John, would have had to find four more sheep for his mother to carry out his father's will (which was the son's legal duty), even if that meant he inherited less than his father intended.

---

86. Swinburne, *A Brief Treatise of Testaments and Last Willes*, p. 292v.

87. Swinburne, *A Brief Treatise of Testaments and Last Willes*, p. 293r.

88. Swinburne, *A Brief Treatise of Testaments and Last Willes*, p. 293r.

89. In the sample dataset, only four of eighty-nine wills (just over 4 percent) named a neighbor (i.e., a nonfamily member) as sole executor.

In this context—whether because a bequest is tied to a particular number or because a general bequest cannot be extinguished—it is actually wise to be unspecific. And so, instead of making a mess of meaning, the collecting phrases used in wills might serve a valuable function. It is, perhaps, their generalizing that makes them useful. If a testator really wanted to ensure that a bequest reached its beneficiary, then a kind of broadness of terminology might sometimes work best. This holds true for all possessions, of course; but it might have a particular meaning when applied to animals, those decaying, changing beings. Recognizing this might help explain why so few wills include specific animal bequests. But recognizing this might also make it possible to see that a will that appears to put cows, pigs, horses, sheep, poultry, geese and bees into general categories, that fails to fully distinguish living beings from non-animate things, and that appears to ignore the animals' sentience and the fact that the testator might have had close, perhaps even affective, relationships with them might actually be doing something very different. It might be ensuring the transfer of ownership and as such attesting to the importance of animals to human lives.

In a world of uncertainty, writing a will that had the highest likelihood of being carried out according to one's intentions was very important. This was good for the testator's soul, good for the testator's family and community, and good for the testator's animals—who would have needed tending throughout the process of death, probate and completion. As Jeremy Cuson, John Clarke, and John Drayne seemed to find when drawing up Widow Sysaye's inventory, counting chickens after they have hatched is difficult but necessary. What one must never do, however, is count one's chickens before they hatch: one must always both expect the unexpected and expect the expected that is uncontrollable: that rye will continue to grow, legatees age and die, animals decay. Such is the messiness of life that the terms "cattle," "chattels," "goods" and "stock" seem to allow room for, that the word "appurtenances" might make possible. These generalizations were not an extinguishing of animals, but were the best way to include them using the limited resources available to the dying. The fact that only 10 percent of the wills from Essex include specific animal bequests should not, then, automatically mean that only 10 percent of people owned and worked with animals or that the dataset offers a glimpse of only 10 percent of the world that included animals. Rather, the wills in general can be interpreted as revealing how complicated living and dying well was in a way that, perhaps, the writings of divines had not intended.

In a ballad that was first printed circa 1680 but was perhaps in circulation well before then, the danger of chicken counting is spelled out in a way that is strangely appropriate to the current discussion:

> I'll tell you a Jest of a Provident Lass.
> Whose Providence prov'd her a Provident Ass;
> She laid forth her store in such brittle Ware,
> That very small profit did fall to her thare;
> Thirteen to the Dozen of Eggs she would buy,
> And set a Hen over them carefully;
> As long as she went her footing she watch'd,
> She counted her Chickens before they were Hatch'd.
>
> Said she, if these Chickens five Capons do prove,
> Capons be Meat which Gentlemen love;
> Those Chickens she would sell to buy a Sow-Pig,
> That it might have young ones e're it was big:
> Then with her Pigs she would have an Ewe,
> It may have Lambs not kill'd with the Dew:
> And as she was thinking to buy her a Calf,
> Her Heels they flew from her a Yard and a half.
>
> Her Heels kiss'd the ground, and up flew her Leggs;
> Down came her Basket, and broke all her Eggs:
> There lay her Pigs, her Chickens, her Lambs
> She could not have young ones except she had Dams:
> Thus Fortune did frown by a fall that she catcht,
> Her Chickens prov'd Addle before they were Hatcht . . .[90]

Raising a stock—to use the terms of the will of Thomas Plumer of Layer-de-la-Haye—required more than wishing.[91] It required foresight, the anticipation of change, and a preparedness—for death as much as for life. The wills of the testators of early modern Essex reveal this and show how far they, like this Provident Lass, foresaw the value of possessing livestock. Unlike her, however, the testators had their chickens before they counted them.

---

90. *The Young-Man & Maidens Fore-cast; Shewing How They Reckon'd their Chickens before they were Hatcht* (London: P. Brooksby, ca. 1680). Christopher Marsh notes that the audience for ballads was potentially popular as well as elite and provincial as well as urban in *Music and Society in Early Modern England* (Cambridge: Cambridge University Press, 2010), pp. 251–54.

91. ERO, D/ACW 10/267.

Keith Wrightson and David Levine read the inventories of Widow Sysaye and her late husband Richard Cesar as presenting "a vivid picture of the living conditions of the laboring poor in the 1620s."[92] Using figures from extant accounts of overseers of the poor from Terling dating from the 1690s, Wrightson and Levine argue that a laborer "would be able to maintain his family at a level slightly above that at which the overseers of the poor maintained the village paupers." A widow would have struggled even more. In this context, and in a village with "no common pasture on which to keep animals,"[93] having a few creatures in the yard—the hog and the poultry—would have made a massive difference to the well-being of Widow Sysaye and her children. A hog would have been fed on household scraps until it would have been slaughtered and salted to provide the household with meat.[94] Poultry would have had longer-term value: eggs were protein for eating as well as a part of medical treatment,[95] and they were also (potential) new hens. For such reasons, the animals we find in wills—the ones specified and the ones potentially included in collecting phrases—would have been valued creatures. Without them, life would have been less nutritionally healthy and it would have been very different, with no daily activities of milking, feeding, bedding down, tending. In addition, we should not underestimate the value of other kinds of engagement—stroking the animals, talking to them—either. The prosaic presence of actual animals—living and breathing, sentient creatures—is something that is all too often excluded from studies but was absolutely central to how people lived, and it is daily life with animals that is the focus of the next chapter.

---

92. Keith Wrightson and David Levine, *Poverty and Piety in an English Village: Terling, 1525–1700* (Oxford: Clarendon, 1995), p. 39. They spell Cesar / Sysaye "Sizer."

93. Wrightson and Levine, *Poverty and Piety*, pp. 40–41.

94. See Robert Malcolmson and Stephanos Mastoris, *The English Pig: A History* (London: Hambledon, 1998), pp. 34–35. It is unlikely that the "little hogge" worth 3s 4d that appeared in Richard Cesar's inventory dated 6 February 1623 was the same "hog," now worth 8s, that appeared in his widow's inventory thirty months later. Even with the different animals, the increase in value of a pig that the two documents include could reflect the fact that Sysaye's inventory was taken in August, so the hog valued then was "further along" than was the one valued in Cesar's inventory in February. ERO, D/ACW 9 / 62 (1623), made probate 66 days after speaking. I return to the agricultural calendar and the slaughter of pigs in chapter 2.

95. See, for example, the various recipes in Gervase Markham's outline of "Houshold Physicke" in his *The English Hovse-Wife, Containing the inward and outward Vertues which ought to be in a compleate Woman* (London: Anne Griffin, 1637), pp. 1–61.

CHAPTER 2

The Fuller Will and the Agricultural Worlds of People and Animals

While a large number of testators used the general terms "goods," "chattels," "cattle," "stock," and "appurtenances" to bequeath their property, there were also those who attempted specificity. The attempts could lead to tediously extended lists of stuff, but these detailed descriptions offer a sense of the range of possessions an individual owned that go beyond what inventories reveal. In addition, where those lists are of animals they also give insight into the perceptions and priorities of the testators and their families and how they worked with and used their quick cattle. This chapter reads the descriptions of livestock in the wills in the large ERO dataset as evidence of attentiveness on the part of the testator and as a means of retrieving some core aspects of life lived alongside animals in the period. It will turn to look at practical husbandry—the centrality of breeding and keeping animals for the people who lived and worked with them—and will also glimpse the possibility of attachment and the interactions that make up a functioning workplace. In detailed testamentary description, I suggest, we can the track worlds.

In early modern Essex, as across the whole country, agricultural animals were living things for use, and they were also individuals the humans knew and with whom they worked. What might seem like an uncomfortable collision of objectification and camaraderie was played out as the norm in the day-to-day lives of many people working with and living with them. Indeed, so

49

"normal" was animal agriculture that many of the nuances of life in the fields and yards of early modern England went without detailed record. There is little discussion, for example, about how people talked to sheep, how they felt when they scratched their pigs, how they knew their cows and knew their cows knew them. But wills, although limited in their interests, can offer ways into worlds that are otherwise difficult to see. A specific example is the best way to begin.

Christopher Fuller was a yeoman from West Bergholt, a village located two miles northwest of Colchester as the crow flies.[1] According to *A History of the County of Essex,* West Bergholt was "one of the least populous parishes in Lexden Hundred in the Middle Ages" and "may have grown rapidly between 1559 and 1601 when baptisms were nearly double the burials, although many people probably migrated to Colchester."[2] Fuller's marriage to Ellis Myll in 1589 is recorded in the parish register of St Mary the Virgin, the village's church, as is his death in 1620. His will is dated 22 June, just seven days before he was interred in the burial ground of the same church, and it was made probate seventy-one days after it had been written, on 1 September.[3]

In his will, Fuller used a number of techniques to identify who was to receive what after his death. Omitting the conventionally Protestant preamble, I give his bequests in full.

> Itm I giue vnto Christopher ffuller my eldest sonne all my teame of horses wth my Cart, wheeles, plough & chaines, horse harnesse both for Cart & plowe & Cart ropes. Itm I giue more vnto him my brended Cowe, my litle black northen Cowe & my Cowe called Shaler, one ~~wennell~~ ^heifer^ of a yeare old one weannell Calfe, ~~and~~ my best sow wth her pigges [piglets⁴] & ~~foure~~ ^two^ shots [young pigs⁵]. Itm I giue vnto him more all my mooueables yt are standing & being in my hall house, one featherbed in the hall chamber, one bolster one pillow, one

---

1. Here and throughout the book I have used the software on "As the Crow Flies" Distance Calculator (http://tjpeiffer.com/crowflies.html) to calculate distance between places.

2. "West Bergholt: Introduction," in *A History of the County of Essex,* vol. 10, *Lexden Hundred (Part) Including Dedham, Earls Colne and Wivenhoe,* edited by Janet Cooper (London: Victoria County History, 2001), pp. 23–27, http://www.british-history.ac.uk/vch/essex/vol10/pp23-27, accessed 14 January 2016.

3. Parish Register of St Mary the Virgin, West Bergholt, ERO, D/P 59/1/1 (1559–1658).

4. Piglet is a term that did not enter usage until the mid-nineteenth century. *Oxford English Dictionary,* s.v. "piglet."

5. See John L. Fisher, *A Medieval Farming Glossary of Latin and English Words Taken Mainly from Essex Records,* 2nd ed., revised by Avril Powell and Raymond Powell (Chelmsford: Essex Record Office, 1997), 40.

Couerlet one paire of blankets & a paire sheets ^a flockbed [bed stuffed with wool]^ and a posted bedstead standing in the hall chamber. Itm I giue him more one powdring [salting] trough in the backe house one brasse pott, one brasse panne two ketles [pans for boiling]. fiue ~~pla~~ pewter platters great & small, one salt seller two ?saucs fiue spoones & one Candlesticke. Itm I giue vnto Willm ffuller my younger sonne my old great brended Cowe two 2 yeare old steares one heifer of a yeare old, two wennell Calfes a sowe wth pigges & two shots. Itm I giue him more one of my old featherbeds wch he will chuse, one bolster, one pillow, ^a paire of sheets & flockbed^ one couerlet, one paire of blanckets & a half-headed bedstead. Itm I giue him more a little old brasse pott & a brazer wheele, the best new ketle & one other ketle. fiue pewter platters a pe[wter] candlestick fiue spoons & one little salt seller. Itm I giue vnto Alce my daughter my great black northen Cowe & one heifer ~~cam~~ that came of ~~her~~ ^my litle^ <northern cowe,> one weannell cow calfe, six pewter platters, six spoones, two litel sauces one ?deep salt seller ^& a litle one^ & one latten [brass-like[6]] candlestick & a litle latten morter & a pestell. A Caldron & a litle ketle, a brasse pott & a litle skellet ~~and~~ a brasse ?barb & a spit. Itm I giue vnto Christopher my son my iron dripping panne & my brasse spitt my least hutch [chest]. Itm I giue vnto Alce my daughter my mooueables standing & being in the parlour, that is to say, one posted bedstead wth a new featherbed new bolster two pillowes one paire of sheets two pillowbers [pillowcases] two blanckets ^my best flockbed^ my best Couerlett one ioyned Cupboard, a table & one ioynt stoole, a danskt chest ~~one~~ ^two^ chaires & all the stained clothes [i.e., colored wall cloths]; and certayne other child-bed linnen being in the dankst chest. Itm I giue to Willm my sone my great skellet. Itm I giue vnto Alce Milles a ~~featherbed~~ ^trundle bed^stead [i.e., bedstead on casters] wth a feather bed a bolster, a paire of blanckets a couerlet a pillow and a paier of sheets. Itm my will is yt my new Cheese made & to be maid till michealmas togethr wth my Corn growing & my hay shall be to the paying of the Rents of my howses land & medow to my landlord & landlady & the ouerplus to be equally diuided betweene my three Children Christopher Willm & Alce. Itm I giue vnto Christopher my son the lease of my howse & ground called Merits. Itm I giue to the poore of the towen of West Bergholt vis viiid of lawfull money of England.

---

6. The *OED* defines latten as "a mixed metal of yellow colour, either identical with, or closely resembling, brass."

And the rest of my moueable goods vnbequeathed I will shall be diuided betwene my sayd three children and I make Christopher & Willm my sons Executors of this my ~~xxxxxxxxxxxxxx~~ ^last will and^ testament.[7]

The page-long will is signed at the bottom with Fuller's mark. On the reverse side it is witnessed by the minister of St Mary the Virgin, Gregory Holland, who was also the scribe of the will,[8] and by Fuller's neighbors, Mathew Damyon, who signed his name,[9] and Godfrey Chamberlin, who signed with a mark.[10]

Christopher Fuller's will is typical in many ways. It comes from an area of Essex where a disproportionately high number of testaments included specific animal bequests. Of the twenty wills from West Bergholt in the large ERO dataset,[11] seven (35 percent) contain such legacies: that is, more than three times the proportion found in the large ERO dataset (where it is 10.1 percent).[12] Of those seven, five include bovids at various stages (but no bulls),[13] four include horses (colts and mares), four sheep (one of which also includes lambs), two include pigs (Fuller's and another will that lists a "sow-hogge"),

---

7. ERO, D/ACW 8/270 (1620), made probate 71 days after writing.

8. Gregory Holland was minister of St Mary the Virgin Church from November 1613 to his death in September 1658. See Parish Register of St Mary the Virgin, ERO, D/P 59/1/1.

9. Mathew Damyon's marriage in 1608 and the baptism of seven of his children are recorded in the parish register of St Mary the Virgin, as is his burial in September 1626. Parish Register of St Mary the Virgin, ERO, D/P 59/1/1. In his brief nuncupative will made on 24 September, Damyon left his son Mathew "my litle black Calfe and nothing els And all that euer I haue beside I giue vnto my wife to bring vp my Children, she shall take all, & pay all." ERO, D/ACW 10/206 (1626), made probate 26 days after speaking.

10. Godfrey Chamberlin is also recorded in the register of St Mary the Virgin. One "Chamberlayne" (no first name) had a son William baptized on 8 January 1609. Marie Chamberlayne (perhaps Godfrey's daughter) was buried on 31 May 1611. A son, Henry, was baptized in April 1613, and another, David, was buried in November 1617. His wife, Ann, was interred on 19 April 1622. There is no further reference to Godfrey Chamberlin after this entry. Parish Register of St Mary the Virgin, ERO, D/P 59/1/1.

11. Two wills in the ERO from West Bergholt are not included in the large ERO dataset: ERO, D/ACW 9/58 (1623) is unsigned and ERO, D/ACW 12/72 has no date for writing the (nuncupative) will, has no probate date, and is damaged.

12. Of the 929 wills from the Archdeaconry of Colchester (ACW), 108 (11.6 percent) include animal bequests. Animal bequests in other will classifications are: Commissary of the Bishop of London (ABW), 9.2 percent (of 2,278 wills); Archdeaconry of Essex (AEW), 12 percent (of 855 wills); Archdeaconry of Middlesex (Essex and Hertfordshire Jurisdiction) (AMW), 8.8 percent (of 352 wills); Archbishop of Canterbury: Peculiar of Deanery of Bocking (APbW), 8 percent (of 25 wills); and Bishop of London: Peculiar of Good Easter (APgW), 0 percent (of 9 wills). The following percentage of wills from parishes within three miles of West Bergholt include animal bequests: Lexden, 0 percent (of 13 wills); Stanway, 0 percent (of 5 wills); Little Horkesley, 8.3 percent (of 12 wills); Marks Tey, 8.3 percent (of 12 wills); Fordham, 17 percent (of 12 wills); and Great Horkesley, 50 percent (of 8 wills).

13. I am avoiding using the term "cattle" in order to prevent confusion of its modern meaning with the early modern usage, which is usually much more general, as I showed in the previous chapter.

and one bequeaths geese and other poultry.[14] It is possible that the high proportion of West Bergholt wills that include animals might be related to the
village's closeness to Colchester, a town that was growing rapidly by the early
seventeenth century.[15] Indeed, if the wills from Colchester in the large ERO
database are anything to go by, residents were highly likely to have looked
outside of the town's boundaries for their meat and dairy products: of the
198 wills from parishes in Colchester, only 6 (3 percent) included animals.[16]
In addition to its closeness to Colchester, West Bergholt lay just under a mile
from the Great Essex Road: thus it was very close to a drover's route, which
meant easy access to meat markets in Ipswich and London.[17]

The style and content of Fuller's will were typical. As with the counting of
chickens in Widow Sysaye's inventory, the oral origin of this written document
is indicated in the movement in Fuller's will between the children—from Christopher to William to Alice, back to Christopher, then back to Alice, to William
again, and then a general bequest. This seems to show a testator remembering things as he goes, to give a picture of a man trying to arrange his affairs in
some hurry. But Fuller's will is also very conventional: it echoes what is found
in many others in the dataset and as such is part of literate legal culture.

---

14. The wills are yeoman Fuller (five cows, three heifers, two steers, four weanels, at least two
horses, two sows, four shots, two groups of piglets), ERO, D/ACW 8/270, made probate 71 days
after writing; the will of a yeoman (one bullock), ERO, D/ACW 9/48 (1623), made probate 24 days
after writing; the will of a widow (three cows, three bullocks, two calves, sheep), ERO, D/ACW 9/61
(1623), made probate 82 days after writing; the will of a widow (two cows, two heifers, one mare,
one colt, one pig, sheep, four geese, poultry), ERO, D/ACW 9/76 (1623), made probate 28 days after
writing; the will of a yeoman (sheep), ERO, D/ACW 9/214 (1625), made probate 91 days after writing; the will of a widow (three cows, one mare, four sheep, five lambs), ERO, D/ACW 10/76 (1625),
made probate 21 days after writing; and the will of a widow (one colt), ERO D/ACW 12/44 (1633),
made probate 84 days after writing. Of the twenty wills from West Bergholt, eight are by widows,
a disproportionately high number (40 percent, compared to 16.5 percent in the large ERO dataset).

15. Figures suggest that the population grew from ca. 3,600 signatories to the oath of allegiance
in 1534 to ca. 11,000 by the time of Fuller's death. See A. P. Baggs, Beryl Board, Philip Crummy, Claude
Dove, Shirley Durgan, N. R. Goose, R. B. Pugh, Pamela Studd, and C. C. Thornton, "Tudor and Stuart
Colchester: Introduction," in A History of the County of Essex, vol. 9, The Borough of Colchester, edited by
Janet Cooper and C. R. Elrington (London: Victoria County History, 1994), pp. 67–76, http://www.
british-history.ac.uk/vch/essex/vol9/pp67-76, accessed 14 January 2016.

16. This low figure for an urban area will be discussed in Chapter 5. Of these six Colchester
wills, one includes horses and carts (ERO, D/ABW 44/5 [1622], made probate after 46 days) and two
include sheep and lambs (ERO, D/ABW 49/185 [1629], made probate 125 days after writing, and ERO,
D/ACW 10/70 [1621], made probate 108 days after writing). The other three Colchester wills with
animal bequests are ERO, D/ABW 49/349 (1627), made probate 69 days after writing, which includes
one bullock; ERO, D/ACW 9/17 (1621), made probate 167 days after writing, which includes "my
cowe & my hogge"; and ERO, D/ACW 10/116 (1625), made probate 72 days after writing, which
included "all my howshold and howseholdstuffe and ymplements of howshold Corne Cowes horses
Carts & furtniture & all other my chattells whatsoeuer."

17. K. J. Bonser, The Drovers: Who They Were and How They Went: An Epic of the English Countryside
(Newton Abbot: Country Book Club, 1972), p. 212.

Like the wills discussed in the previous chapter, Fuller attempted to organize his world, and his moveable goods that are not self-moving fall into three key categories: furniture (beds and associated bedding play a particularly significant part, as is frequently the case in wills from this period); household furnishings and utensils; and equipment for preparing food and cooking. Fuller did not distribute clothing (usually termed "apparel" in this period), although it was typically included in wills:[18] the "stained clothes" he did bequeath were likely to be colored (not discolored) wall hangings. Aside from the lack of apparel, Fuller's testament resembles numerous others. The detailed listing of his household possessions—the flockbeds, featherbeds, and trundle bed; the best new, the little, and the other kettles; the pewter candlestick and the latten one—enabled bequests to be understood and, one imagines, correctly distributed. Moving outside the house, Fuller's will also contained evidence of agricultural production. The cheeses made and about to be made reveal the work of a dairy, and the corn signals arable farming.

Fuller names few legatees—his three children and "Alce Milles" who, it is possible, was his niece by marriage and/or his goddaughter.[19] The small number of people who received bequests reflects something that Keith Wrightson and David Levine found in their study of the village of Terling (located about fifteen miles as the crow flies from Fuller's home in West Bergholt): early modern testators tended to bequeath their goods to their closest relatives. Wrightson and Levine note of the 192 wills in their sample that "mention of kin beyond children (140 wills) and spouse (116 wills) was rare."[20] Fuller's wife Ellis (a likely early modern spelling of Alice) had died in February 1618 (her burial was recorded in the Parish Register of St Mary the Virgin), but his status as a widower might have been guessed at without that external evidence because of her absence in his will: existing relatives who were to be excluded from inheritance often got a small bequest to signify that they had not simply been forgotten.[21]

---

18. On bequeathing clothing in the period, see Ann Rosalind Jones and Peter Stallybrass, *Renaissance Clothing and the Materials of Memory* (Cambridge: Cambridge University Press, 2000).

19. Fuller's wife's maiden name was Ellis [Alice] Myll. Another "Alse Mills" (perhaps the one in Fuller's will) was baptized in St Mary the Virgin church in West Bergholt on 3 February 1611. Parish Register of St Mary the Virgin, ERO, D/P 59/1/1.

20. Keith Wrightson and David Levine, *Poverty and Piety in an English Village: Terling, 1525–1700* (Oxford: Oxford University Press, 1995), p. 92.

21. Ralph Houlbrooke notes this, referring to a case from 1745. See Houlbrooke, *Death, Religion, and the Family in England, 1480–1750* (Oxford: Oxford University Press, 1998), p. 94.

Fuller passed on the lease of Merits, the only property referred to in the will, to his oldest son Christopher, and we can see from this that Fuller is recognizing him as his key legatee.[22] Indeed, according to a survey of West Bergholt taken one year later in September 1621, Christopher Fuller junior, in possession of almost fifty acres of leasehold, was the second largest land holder in the village. The two witnesses to Fuller senior's will are also listed in this survey: Mathew Damyon had twenty-four acres of leasehold and Godfrey Chamberlin had ten acres of copyhold land.[23] Given this, it is perhaps unsurprising that unlike Sysaye, the Terling widow, Fuller was able to bestow wealth in goods upon all of his children. This seems to challenge in part what Cicely Howell noted of contemporary inheritance customs in the Midlands, where, she writes, "We learn from the wills that the son who received the land did not receive the livestock and gear: these went to the remaining children. The heir could then either buy out his brothers and sisters, or could continue to support them on the tenement in return for the use of the livestock and the gear, at any rate for the first few years."[24] It is possible that Fuller's children could have shared the animals, but the difference between Howell's findings and Fuller's will (the fact that Christopher got the property and also inherited a sizable proportion of the livestock) might be for a number of reasons: perhaps all of Fuller's children had already received some land from their father before June 1620 (say, at their marriages or on their mother's death), maybe they were already well established in their own homes by the time of his death so sharing the livestock was not necessary.[25]

---

22. The parish register for St Mary the Virgin in West Bergholt records that "Goodman ffuller had a sonne Cristen'd" on 29 April 1593 and that Alice "the dawighter of Cristoffer fuller" was christened on 27 February 1596. No second son is recorded in the register, and Christopher, the oldest son (who may or may not be the child born in 1593), never appears there by name. William, as is shown below, was recorded as marrying and having multiple children. Parish Register of St Mary the Virgin, ERO, D/P 59/1/1.

23. "A survey taken in September last 1621 of the lands of Sr John Denham knight In parishes of Barfold ffordam and lextin," ERO, D/DMa M18 (1621).

24. Cicely Howell, "Peasant Inheritance Customs in the Midlands, 1280–1700," in *Family and Inheritance: Rural Society in Western Europe, 1200–1800*, edited by Jack Goody, Joan Thirsk, and E. P. Thompson (Cambridge: Cambridge University Press, 1976), p. 152.

25. The parish register for West Bergholt shows that William Fuller, Christopher's younger son, married his wife Joan in March 1617 and that they had a child, Alce, baptized on 8 March 1618. Another child, William, was baptized on 4 September 1620 and was buried nine days later. Another son—William again—was baptized on 12 August 1621 and was buried three months later. Yet another son, Christopher, was baptized on 9 February 1623. The next record of William Fuller senior is in January 1628: "of Will. ffuller & Joane his wife bapt." The parson, Gregory Holland, did not include the child's name or sex. Fuller's wife was buried on 15 February 1631. There is no further record of William after this. Parish Register of St Mary the Virgin, West Bergholt, ERO, D/P 59/1/1.

Fuller called himself a yeoman, a term with a broad meaning in this period. However, Howell's study suggests that more might be evident in the will's bequests. She writes:

> The hard core of small property holders is conveniently and clearly revealed during the period of high agricultural prices after 1636. They group themselves at one end of the spectrum as leaving legacies predominantly in kind, and they were joined as the century wore on by a proportion of the middle group who had previously left legacies of £10–20 to each of their children. At the other end of the spectrum one can see a growing number of larger property holders, men who could leave £200 or more to their children. The number at the centre grows thinner as the process of polarization continues. The parting of the way between husbandman and yeoman can be seen taking place, the one to remain the humble villager, the other to become the local squire.[26]

Although Fuller's will was written more than a decade before the key shift Howell recognized and from a geographical location at some distance from the wills from Leicestershire that she studied, it is possible to read its bequests as signaling that even though they do not include ready money, Fuller's life on Merits was well above subsistence farming. The inclusion of numerous bovids at various stages and ages signals his possession of a herd large enough to support calves and fulfill human requirements (providing enough milk to make cheese). Fuller was certainly not one of the laboring poor, as Wrightson and Levine termed widow Sysaye.[27]

The value and nature of the stuff included in the collecting phrase at the end of Fuller's will—"And the rest of my moueable goods vnbequeathed"— should not, of course, be underestimated, but whether it is a comprehensive representation of his property or not, what the will's detailed contents make visible was his care for his children. Such planning was typical: indeed, supporting the well-being of one's relatives with bequests of animals could start years in advance of their independence. Thomas Plumer of Layer de

---

26. Howell, "Peasant Inheritance Customs in the Midlands," p. 152.

27. Wrightson and Levine, *Poverty and Piety in an English Village*, p. 39. In the large ERO dataset, 95 percent of yeomen included monetary bequests in their wills. Of the twenty yeomen's wills in the sample dataset, only one did not include money, and of the nineteen that include monetary bequests, nine included £10 or less. In the large ERO dataset, 82.5 percent of husbandmen included monetary bequests. Of the twelve husbandmen's wills in the sample, two did not include any money and seven included £10 or less. There are only four yeomen from West Bergholt in the large ERO database. Fuller was the only one of them to not bequeath money; thirteen of twenty (65 percent) of the village's testators left money as legacies.

la Haye (which is four miles from West Bergholt), for example, left "vnto Willam Scarlet my Granchilde one Cowe bullocke of two yeares owlde to raise him a stocke" in his 1627 will.[28] From one young animal might grow a yard full (I return to the term "cow bullock" later in the chapter). In addition, bequests to young legatees were often held by the executor until the child reached maturity—usually at age 21, but occasionally older[29]—or until the day he or she married, whichever came first. Thus, Thomas Judd, a husbandman from Great Canfield, bequeathed to "Sarry my daughter one Cowe to be deliured at the day of death of Katheren my wife or at hir day of marage the wch shall happen first."[30] Such bequests ensured that the legatee would have something to take with them as they established their own household.

Even though Fuller referred to himself as a yeoman, he could have had difficult choices to make to secure his children's well-being. He had clearly achieved a level of security: the fact that the corn, hay, and cheeses might be sold (or exchanged) for rent tells us that his household produced more than it might use itself, and the presence of young animals signals that he was likely to be making money from selling them on. These possessions, along with Howell's analysis, underline Fuller's status as a self-sufficient farmer. But the will reveals more than this. As Wrightson and Levine note of the detailed nature of the contents of wills written during the sixteenth and seventeenth centuries:

> These matters are not trivial. These various goods were their owners' pride, lovingly detailed in will after will. The fine table laid with pewter and brass, the great bedstead in the parlor with its featherbed, pillows, and coverlet . . . were the outward and very visible signs of success among those best placed to take advantage of the opportunities of the age. The prestige that they conferred was perhaps the more important

---

28. ERO, D/ACW 10/267 (1627), made probate 98 days after writing. Money might also be bequeathed with the specific intention that it would be used to buy animals for younger legatees. For example, John Lukin of Pleshey gave his two younger daughters "al my readie Moneye . . . to be imy imployed to ther benefit to raise a stocke for them." ERO, D/ABW 44/211 (1622), made probate 98 days after writing. In another example, George Gouldringe, a husbandman from Lambourne, proposed that money should be "put forthe and be to To run in stocke for [his grandchildren] vntill they shall come vnto one an Twenty yeares of age." ERO, D/AEW 19/113 (1631), made probate 327 days after writing. See also ERO, D/AEW 17/30 (1621), made probate 123 days after writing. I return to the frequent bequeathing of animals to children in chapter 4.

29. John Waight of St Leonard, Colchester, for example, bequeathed his son Thomas £13 6s 8d "when hee shalll accomplishe his full age of foure and Twenty yeares." ERO, D/ABW 44/5 (1622), made probate 46 days after writing.

30. ERO, D/ABW 48/27 (1627), made probate 978 days after writing.

because these comforts were new and because they were not usually available to the mass of the villagers.[31]

Fuller was passing on to his children not simply pewter platters and beds but the status and security he had achieved. And that status and security was heavily indebted to his quick cattle.

Fuller's will includes more than twenty animals (perhaps more than thirty): the team of horses (likely four animals, but maybe more[32]), five cows, two steers (young castrated male oxen), three heifers (young cows), four weanels (newly weaned calves), two sows, four shots (young pigs), and two litters of piglets. These animals were productive and known members of the household, but all of them—named or otherwise (I come to Shaler in the next chapter)—were property. This does not relegate the animals to serving as an inanimate background to the family's story, however.[33] Rather, because animals were—as the inventories showed—among the most valuable things a testator might possess, as well as the most productive (as is evident in that bequest of a young cow to a grandson to "raise him a stocke"), they also played a crucial role in the testator's plans for their family's future. Counting one's chickens was, once again, vitally important.

This reliance on animals is something that is evidenced by the fictional representation of a poor man's attempts to reclaim from a usurer the cow he has put down as surety for a loan in Thomas Lodge and Robert Greene's 1590 play *A Looking Glass for London and England*. The usurer tells the poor man that he has missed the day of reclamation and thus "thou getst no Cow at my hand." The poor man's response reveals the

---

31. Wrightson and Levine, *Poverty and Piety in an English Village*, p. 39.

32. According to Markham, "the number fit for the plow is eight, sixe, or foure; for the Cart, fiue or foure." Gervase Markham, *Markhams Farewell to Hvsbandry* (London: I. B. for Roger Jackson, 1620), p. 148. In his will of 1623, John Smith of Hempstead bequeathed his son John "my Carte with all my furnitierie thervnto belonginge with fiue horses." ERO, D/ABW 44/308 (1623), made probate 113 days after writing.

33. Virginia DeJohn Anderson notes that in the narrative of early modern colonization, livestock "usually serve as part of the scenery rather than as historical actors." Her study of the impact of the transfer of English livestock farming to the New World challenges this conception brilliantly. Anderson, *Creatures of Empire: How Domestic Animals Transformed Early America* (Oxford and New York: Oxford University Press, 2004), p. 1. For an excellent study of the impact of cows on another continent, see Sandra Swart, "Settler Stock: Animals and Power in Mid-Seventeenth-Century Contact at the Cape, circa 1652–62," in *Animals and Early Modern Identity*, edited by Pia F. Cuneo (Farnham: Ashgate, 2014), pp. 243–67.

importance of the animal to him and his family, an importance that goes beyond its monetary value:

> No Cow sir, alas, that word, no Cow, goes as cold to my heart as a draught of small drinke in a frostie morning. No cow sir why alas, alas, M. Vsurer, what shall become of mee, my Wife, and my poore child? . . . Why sir, alas, my Cow is a common wealth to mee, for first sir, she allowes me, my wife and son, for to banket [banquet] our selves withal, butter, cheese, whay, curds, creame, sod [boiled] milke, raw-milk, sower milke, sweet milke, and butter milke, besides sir, she saued me euery yeare a peny in almanackes, for she was as good to mee as a Prognostication, if she had but set vp her taile, and haue gallapt about the meade, my little boy was able to say, oh father, there will be a storme: her very tayle was a Kalender to mee, and now to loose my Cow, alas M. Vsurer, take pitty vpon me.[34]

For the poor man, the cow is central to his family's well-being. She is a provider of food and drink that was vital to a healthy diet. In this period, Craig Muldrew has argued, the regular inclusion of dairy products in meals would have made it likely that the eater might earn more because he or she was capable of performing better at physical labor than others who ate less nutritious food (Muldrew's example is scything).[35] In addition, the poor man's cow can predict the weather, a valuable capacity for people working the land. Natural meteorologists, animals were trusted to know the processes of nature in a way that humans alone could not.

Lodge and Greene's representation of the poor man's sense of loss was performed in front of a London theater audience, and that may have impacted the scene's being played for comedy (I come to contemporary Londoners' growing distance from quick cattle in chapter 5). However, the devastating potential of the loss of an animal was not just something that was staged in the capital. It can also be found in the real world in Keith Wrightson's reference to a help ale "held by a Wiltshire villager who had lost his only cow and was advised by a neighbour 'to provide a stand of [ale] and he would bring him company to help him towardes his losses.'"[36] Here, the struggling individual, desperate to replace his cow, sold the ale he provided to his neighbors at a profit to raise funds to buy a much-needed new animal. Judith M. Bennett has noted the

---

34. Thomas Lodge and Robert Greene, *A Looking Glass for London and England* (London: Barnard Alsop, 1617), B4r.

35. Craig Muldrew, *Food, Energy and the Creation of Industriousness: Work and Material Culture in Agrarian England, 1550–1780* (Cambridge: Cambridge University Press, 2011).

36. Quoted in Keith Wrightson, *Earthly Necessities: Economic Lives in Early Modern Britain, 1470–1750* (2000; repr., London: Penguin, 2002), p. 78.

potentially coercive aspects of such church ales,[37] but in 1592 Robert Greene used one to mark the continuity of community. In *A Quip for An Upstart Courtier*, the Italianate upstart Velvet Breeches and his homely challenger Cloth-breeches come upon "a mad merry crew" led by "a plain country Sir John, or vicar[,] that proclaimed by the redness of his nose he did oftener go to the alehouse than the pulpit." This vicar defends himself by saying that "these all are my parishioners, and we have been drinking with a poor man and spending our money with him, a neighbour of ours that hath lost a cow."[38]

The fact that human wellbeing would suffer if animals were lost or sick or died was a key motivation for testators when writing their wills: ensuring that a family member received an animal might be ensuring that person's continuing security. But the value of animals was also a key motivation for testators' attentiveness to the health of their quick cattle. Careful consideration was necessary. Thomas Tusser, writing in verse, noted, "If Hog doe cry, / give eare and eye."[39] In the same vein, and also in verse, Leonard Mascall wrote:

> Thy horse once sicke deferre no time
> his griefe for to appease:
> For sorenesse oft so dangerous is,
> thou maist thy horse soone leese.
> If thou doest marke of sorenesse most,
> whereof they doe proceed,
> Thou shalt finde out that most doe come
> for lacke of taking heede.[40]

And, referring to an animal that does not appear in any Essex will, Gervase Markham wrote, "If you perceiue your Goates to droope, or looke with sullen or sad countenances, it is an assured signe of sicknesse."[41] By implication, good

---

37. Judith M. Bennett, "Conviviality and Charity in Medieval and Early Modern England," *Past and Present* 134 (February 1992): 19–41.

38. Robert Greene, *A Quip for An Upstart Courtier: Or, A quaint dialogue between Velvet breeches and Clothbreeches* (London: Iohn Wolfe, 1592), G2r–v.

39. Thomas Tusser, *Five Hundred Points of Good Husbandry* (London: I. O., 1638), p. 27.

40. Leonard Mascall, *The Government of Cattle* (London: Tho. Purfoot, 1627), p. 99. For a survey of the emerging printed genre of animal healthcare manuals in the period, see Louise Hill Curth, *The Care of Brute Beasts: A Social and Cultural History of Veterinary Medicine in Early Modern England* (Leiden: Brill, 2010). Mascall and Tusser may have used verse to make their work more memorable to those who had it read to them instead of reading it themselves.

41. Gervase Markham, *Cheape and Good Hvsbandry: For the well-Ordering of all Beasts, and Fowles, and for the generall Cure of their Diseases* (London: T. S. for Roger Jackson, 1616), p. 119. Miranda Chaytor found no goats in the probate inventories of early modern Ryton, County Durham, but she did find them recorded in the manorial court records. She argues that this suggests that "goats . . . seem to have been kept only by immigrant labourers' households." Chaytor, "Household and Kinship in Ryton in the Late 16th and Early 17th Centuries," *History Workshop Journal* 10, no. 1 (1980): 32.

owners should study to know their animals in health so that they might recognize quickly when something was wrong. Paying close attention was vital.

The modern agricultural ethicist Bernard E. Rollin has termed this self-serving concern with animal health "the 'social contract' between humans and animals." In this contract, reciprocity was key: humans protected and fed their animals and animals worked and provided food for their humans. Rollin argues that the social contract was dominant in farming until the mid-twentieth century, when the "huge scale of industrialized agricultural operations" took over. For him, pre-industrialized farming had as its logical corollary an "anti-cruelty ethic": not needing legislation to guide them, agriculturalists in the early modern period, so this reading goes, policed themselves in their treatment of their animals, and kindness was the predominant attitude because it was the most productive.[42] This was likely to be true for many who worked with animals, and thus, although criminal law did not address animal welfare until the nineteenth century, early modern moral law recognized the importance of animal well-being. Writing in 1612, and inspired by Thomas Aquinas, the clergyman John Rawlinson, for example, voiced something that was akin to Rollin's idea of the "anti-cruelty ethic" when he wrote: "As beasts must not be *over-burden'd*; no more are those that haue *milk*, to be *over-milkt*. For . . . he that brings his beast too oft to the milk pale; shall in the end, in sted of milke, haue nothing but bloud."[43] Here the larger focus is on humans rather than animals (on the danger to the dairy of over-milking, rather than on the harm caused to the cow). But even such self-centered ethics had their limits, and while evidence from early modern animal bone remains reveals care (for example, a pig's leg bone with evidence of a break that had been treated), it also reveals carelessness. Some creatures were clearly viewed as expendable and were worked while in curable pain because, it must be assumed, the pain did not interrupt the animal's capacity to work. The zooarchaeologist Richard Thomas, for example, has found bones dumped at the end of the seventeenth century at Dudley Castle that show a horse suffering from a massive bone infection in its leg caused by an injury. There is no evidence of treatment on the bone.[44]

Rollin's rather Smithian reading of pre-industrial human-animal relations (high productivity from unregulated producers is caused by and thus produces

---

42. Bernard E. Rollin, *Farm Animal Welfare: Social, Bioethical, and Research Issues* (Ames: Iowa State University Press, 1995), pp. 5–6, 10, and 8.

43. John Rawlinson, *Mercy to a Beast* (Oxford: Joseph Barnes, 1612), p. 26.

44. See Richard Thomas, "Diachronic Trends in Lower Limb Pathologies in Later Medieval and Post-Medieval Cattle from Britain," in *Limping Together through the Ages: Joint Afflictions and Bone Infections*, edited by G. Grupe, G. McGlynn, and J. Peters (Leidorf: Rahden/Westf., 2008), pp. 187–201. See also Erica Fudge and Richard Thomas, "Visiting Your Troops of Cattle," *History Today* 62, no. 12 (2012): 41.

good welfare) is not the only way of viewing early modern agriculture, however. There is another reason for assuming that the testators and legatees of early modern Essex would have paid close attention to their animals that is, perhaps, almost too mundane to be recognized as ethical at all. Quick cattle were sentient and self-moving beings, and how to live with and work with them—working out who they were, if you like—was crucial to the functioning of the household. Getting to know one's livestock was a vital part of a family's survival, and getting to know them would have been particularly easy on a smallholding where few animals were kept. In such an environment, in addition to looking at their outsides, the person who worked most closely with the animal would also have a grasp of an animal's interior being—of what might be termed their character.[45] Understanding which side a cow liked to be milked on, for example, could save you from a kicking. Only a person who took the cow's individual preferences seriously would be aware of this and act upon that awareness.[46]

I return to the limits of the agricultural social contract in chapter 4, but we should note here that whatever care was given to quick cattle, the relationship between humans and animals was always one-sided. As now, in the early seventeenth century no pig could plead for its life come the time of its slaughter, no cow could walk away from an abusive owner. And if an animal did try—if a horse refused to pull the plow—it would be regarded as unusable as anything but hide to be sold to a tanner and as flesh to be fed to the dogs. Animals were expected to behave in certain ways (although minor idiosyncrasies were allowed for), and those who failed to do so would be removed and destroyed.[47]

---

45. Indeed, the outer and the inner were sometimes linked. Gervase Markham, for example, argued that the color of a horse had implications for its temperament, that it was related to the animal's humoral makeup: "The Horse that is pure blacke, and hath no white at all vppon him, is furious, dogged, full of mischiefe and mis-fortunes." Markham, *Cavelarice, Or the English Horseman* (London: Edward Allde, 1607), "The Seconde Booke," p. 6.

46. On the outcome of failing to milk a "skittish" cow on the right side, see the contemporary ballad "The woman to the plovv and the man to the hen-roost; or, A fine way to cure a cot-quean," ascribed to M. P. [Martin Parker] (London: F. Grove, 1629). I discuss this in "Farmyard Choreographies in Early Modern England," in *Renaissance Posthumanism*, edited by Joseph Campana and Scott Maisano (New York: Fordham University Press, 2016), pp. 145–66.

47. An extreme version of this can be traced in 1682, when the Earl of Rochester's vicious horse was baited to death at the Bear Garden in London. See *Loyal Protestant and True Domestick Intelligence*, nos. 141 and 142 (13 April 1682 and 15 April 1682). On a more mundane level, a 1630 case brought before the Consistory Court of London (the court that investigated moral crimes) recorded—as part of a case focusing on the insulting of one of the (human) parties—the fury of a butcher who had bought a cow that "provd like a bitch." Carpenter *c.* Turner, Consistory Court of London, Deposition Book LMA, DL/C 233 (1630), fols. 161r–v. One implication is that the cow was likely to end up on the slab because of her character failings. Laura Gowing, however, offers another possibility: a witness in a later deposition from 1658 used the term "bitch" to mean an infertile woman, so the 1630 complaint could be about the animal's lack of productivity. Ireland *c.* Francis Powell, DL/C 235 (1658), cited in Gowing, *Domestic Dangers: Women, Words and Sex in Early Modern London* (Oxford: Oxford University Press, 1996), p. 76.

The requirement that animals fulfill their function remained even when relationships were long established. Old animals who had served their owners over a number of years were viewed as expendable: zooarchaeological findings reveal that it is typical to find "large numbers of lower limb bones" from bovine animals with butchery marks on them in late medieval castle sites. These bones, Thomas writes, were from mainly old creatures, many of whom would be "breeding animals"; that is, dairy cows.[48] This shift from milker to meat was simply the passage of her life. In her 1620 will, for example, Mary Archer, a widow from Little Clacton, seemed to voice a lack of care: "I will that myne executor tak in my two old cowes at Micholmas that are now in the possession of Henrie Soyars and that he shall sell them and putt out two young Cowes in sted of them to the vse and benefitt of John Archer and Robbecah Archer the two youngest children of my sonne Thomas Archer."[49] There was only one likely buyer for "two old cowes." It is tempting to regard Archer as unsentimental, but that would, I think, be anachronistic. The request might, in fact, be read as a record of both affection and pragmatism. Affection because Archer had not wished to part with the old animals during her own lifetime,[50] but also pragmatism because when they were no longer useful, quick cattle were no longer kept because keeping animals was expensive in food and time: up to 20 percent of arable production went on their upkeep.[51] In *The Whole Art and Trade of Hvsbandry*, Conrad Heresbach wrote, "You must euery yeere . . . sort your stocke [i.e., dairy cows], that the old that be barraine, or vnmeete for breeding, may be put away, sold, or remoued to the Plow: for when they be barraine (as *Columella* saith) they will labour as well as Oxen, by reason that they are dried vp, but we vse commonly to fat them."[52] Indeed, In the "most pleasant and merie new Comedie" *A Knacke to knowe a Knaue*, the Farmer reveals that the killing of ageing cows was so expected that it had taken on proverbial value:

As the tallest Ash is cut down, because it yeelds no fruit,
And an vnprofitable cow, yeedling no milke, is slaughtred,

---

48. Thomas, "Diachronic Trends in Lower Limb Pathologies," p. 188.

49. ERO, D/ABW 43/197 (1620), made probate 107 days after writing. Other wills also include directions to sell animals. See ERO, D/ABW 44/5 (1622), made probate 46 days after writing; ERO, D/ABW 44/308 (1623), made probate 113 days after writing; and ERO, D/ACW 10/19 (1626), made probate within two months of speaking.

50. This is something I return to in chapter 5.

51. Mark Overton, *Agricultural Revolution in England: The Transformation of the Agrarian Economy 1500–1850* (Cambridge: Cambridge University Press, 1996), p. 14.

52. Conrad Heresbach, *The Whole Art and Trade of Hvsbandry Contained In foure Bookes* (London: T. S., 1614), p. 121v. Leonard Mascall likewise recognized that barren cows could be useful as meat or as draft animals: "yee must either fat them, or occupie them to the plough," he advised. Mascall, *Gouernment of Cattle*, p. 58.

And the idle Drone, gathering no honie, is contemned,
So vngrateful children, that will yeeld no naturall obedience,
Must be cut off, as vnfit to beare the name Christians.[53]

When the animals fulfilled their allotted roles, care was certainly a practical response, but when they ceased to be productive for the household or when they reached the appropriate weight or size for sale, that response changed. There was, thus, a paradox at the heart of husbandry: vigilant attention was required, but that vigilance would always end in the premature death of the animal. I return to this in chapter 4.

Despite the fact that the shadow of slaughter always hung over husbandry, careful and care-filled attention to the well-being of animals was the cornerstone of people's relationship with their quick cattle.[54] How that vigilance manifested itself, however, is not something we can be sure about as there may have been more to it than looking out for ailments or checking for sad countenances. Good husbandry might best be found in actions that are difficult to trace as they have left little record because they were so ordinary. Having a sense of how people talked to, tickled, patted, worried about, fussed over their animals might also be crucial to having a sense of how they—humans and animals—lived. And they lived together.

It is worth understanding the work that had to be done when living on a farm to get a sense of the centrality of animals to the lives of many people in the early seventeenth century. According to Gervase Markham, at the start of the calendar year the husbandman's day would begin at "foure of the clock in the morning" with prayers, after which the first task was to tend the livestock.

> he shall goe into his stable, or beast house, and first he shall fodder his cattell, then cleanse the house, and make the boothes neate and cleane, rubbe downe the cattell, and cleanse their skinnes from all filth, then he shall curry [comb] his horses, rubbe them with cloathes and wispes, and make both them and the stable as cleane as may be, then he shall water both his Oxen and Horses, and housing them againe, giue them more fodder.

---

53. Anonymous, *A Knacke to knowe a Knaue* (London: Richard Iones, 1594), B4r. Francis Meres used the same sequence of natural events and their consequences, although his final one concerns not ungrateful children but "the women that maketh her selfe barren by not marrying." She is "accounted . . . worse then a carrion, as *Homer* reporteth." Meres, *Wits common wealth The second part* (London: William Stansby, 1634), p. 96.

54. See Fudge, "Farmyard Choreographies," pp. 146–47. Being careful, I suggest there, might be a professional obligation, whereas being "care-filled" is a more focused, personal attention.

Even as the animals were eating, the husbandman "shall make ready his col-
lars, harnes, treates, halters . . . seeing euery thing fit" for the plow. This would
all, Markham wrote, take two hours, after which the husbandman could go
to his own breakfast. After that the "gearing and yoaking" of the animals
would take another half hour. Only then could plowing could begin, and that
would go on "from seauen of the clocke in the morning, till betwixt two and
three in the afternoone [when] he shall vnyoake and bring home his cattell,
and hauing rub'd them, drest them, and cleansed away all durt and filth, he
shall fodder them and giue them meat." The husbandman then received his
own dinner (taking half an hour over it), by which time it was four o'clock
and time to "goe to his cattell againe" to rub them down some more, cleanse
their stalls, give them more fodder, and "make ready fodder of all kinds, for
the next day." Then "he shall goe water his cattell and giue them more meat"
and at six o'clock "he shall come in to supper." After this, Markham says, the
husbandman would sit by the fire and "doe some husbandly office within
dores," such as mending shoes or stamping apples for cider, before finally,
at eight o'clock, taking "his lanthorne and candle," he would go back "to his
cattell, and hauing cleansed the stalles and planckes, litter them downe, looke
that they be safely tyed, and then fodder & giue them meate for all night"
before "giuing God thankes for benefits receiued that day" and going to "rest
till the next morning."[55] While the time spent caring for livestock—cleaning,
housing, feeding them—might change once spring arrived and the animals
were put out to pasture, even if Markham exaggerates this aspect of the husbandman's day over the winter months, his outline serves as a useful reminder
of just how much attention they required. As Anderson has noted, "The ideal
husbandman spent far more time each day with his livestock than with his
wife and children."[56]

The farmer's wife also had an arduous life. Where the husbandman's
"office and implyments are ever for the most part abroad, or removed from
the house, as in the field or yard," Markham wrote, the "Huswife . . . hath
her most generall implyments within the house." Although innuendo was
the object of the account, Scogin, Andrew Boorde's satirical antihero, accidentally gave a useful outline of a woman's work when he wrote a letter in a
comically failed defense of his local priest:

After commendations, I certifie you, that where my Priest is complained
on for a woman that he keepeth in his house, to wash his dishes and to

55. Markham, *Markhams Farewell to Hvsbandry*, pp. 145–46.
56. Anderson, *Creatures of Empire*, p. 85.

gather rishes, to milke his cow & to serue his sow, to feed his hen & cocke, to wash shirt and smocke, his points to vnloose, & to wipe his shooes: to make bread & ale, both good, & eke stale, & to make his bed, & to looke his head, his garden she doth weed, & doth helpe him at need: no man can say, but night and day, he could not misse to clip & kisse: she is faire and fat, what for all that, I can no more tell, but now fare you well.[57]

Markham, with more seriousness, noted that the housewife was responsible for the behavior and health of the family (and thus needed an understanding of religion and of herbs and treatments for illness and injury); for tending the kitchen garden; and for cooking, brewing, and spinning. In households with cows, she was also responsible for the work of the dairy, which included milking and tending cows, weaning calves, and making butter and cheese (in a world without constant refrigeration, milk needed to be rapidly transformed into forms that remained edible for longer). Dairying would take up much of the housewife's time. In addition to doing the actual milking, the wife had to keep the dairy clean: "not the least moate of any filth may by any meanes, appeare, but all things either to the eye or nose so void of sowernesse or sluttishnesse, that a Princes bed-chamber must not exceede it," Markham wrote. This would include the "delicate keeping of her milk-vessells." These tasks were necessary whether the household had one or forty cows; and this is why small-scale dairying was particularly labor intensive.[58]

The actual work involved in making butter included straining, skimming, and churning milk, then tossing the newly churned butter in water; checking it for hairs, and potting and salting it. For cheese making the housewife would, at the beginning of the year, prepare *"Runnet, which is the stomacke-bagge of a yong suckling calfe which never tasted other foode than milke, where the curd lieth undigested."* Some of the runnet (now called rennet) would be added to

---

57. Andrew Boorde, *The first and best part of Scoggins iests* (London: Francis Williams, 1626), p. 23.

58. Gervase Markham, *The English Hovse-Wife* (London: Anne Griffin, 1637), pp. 1–2, 194–95; Muldrew, *Food, Energy and the Creation of Industriousness*, p. 235. Wills that include dairy cows in the large ERO dataset give a sense of three cows on average. This figure is an underestimation because it is calculated by assuming that phrases such as "a cow," "my cows," or "some cows" mean that the testator possessed just two animals if no further specificity was offered. Some wills mentioned much larger numbers. For example, one will mentioned forty cows (ERO, D/ABW 43/178 [1620], made probate 1465 days after writing); another mentioned twenty-two cows (ERO, D/ABW 44/173 [1622], made probate 153 days after writing); and another eighteen cows (ERO, D/ABW 52/185 [1635], made probate 51 days after writing). In his study of wills and inventories from Derbyshire, D. J. Baker found that "the average herd before 1641 numbered . . . four cows." Baker, "Stanton-by-Bridge: A Study of Its People from Wills and Inventories, 1537–1755," *Derbyshire Miscellany* 10, no. 3 (1984): 81.

the new milk, and when curds formed, the housewife would mash and break them by hand, remove the whey, turn and strain the cheese, press it, dry it, press it again, and then salt it and lay it up.[59] On top of this, the wife oversaw the poultry[60] and might keep an eye on any pigs in the yard (they were often fed on waste skimmed milk from the dairy, among other things, and shared the chaff from threshing with the hens[61]). During harvest time, she would also work with her household in the fields.[62] In *A dairie Booke for good huswifes* (1588), Bartholomew Dowe's "Suffolcke man" states that his mother and her maids were "euery daie in the yeere Winter and Sommer vp out of their beddes, before foure of the clocke euery morning."[63] And Markham proposed that the housewife's day might begin, "betwixt five and six in the morning" with the first milk of the day and end, like her husband's, at about nine or ten o'clock at night.[64]

For testators whose engagement with agriculture was a second job—those who listed themselves as weavers, blacksmiths, and carpenters but who owned livestock, for example—their work with their animals would fit around their other occupation. Thus, for example, the will of the East Donyland mariner Samuel Parish reveals substantial productivity on every level: he bequeathed his "customary messuage [that is, land that had, or could have, a dwelling house on it] or Tenemt wth all the Lands tenemts hereditamts and appurtenancs to the same belonginge" to his wife Alice, which, after her death, was to go to his son John. Other messuages were left to each of his sons Samuel and Edmond, and to his daughter Anne, while another daughter, Rachell, got two "freehold Tenemts wth thappurtenancs." The will mentions "the cropp of corne ~~gr~~ hopps & fruite growing vppon the Land" of one of these tenements, showing arable farming being carried out, and in addition, his wife got "three cowes [and] six sheepe," "two of my best pigges," and "a quarter of ^the^ Cheese in my howse," revealing the work of a dairy, participation in the wool trade and, perhaps, meat production. On top of this evidence of

---

59. Markham, *English Hovse-Wife*, Chapter 6 passim. The quote about dairying is on p. 201.

60. It is notable that while Markham did not mention poultry in his outline of the housewife's labors, Leonard Mascall dedicated *The Husbandlye ordring and Gouernmente of Poultrie* (London: Thomas Purfoote, 1581) to a woman, despite the gendering of the book's title.

61. Tusser wrote: "Keepe threshing for thresher, till May be come in, / to haue to be sure, fresh chaffe in thy bin: / And somewhat to scamble, for Hog and for Hen, / and worke when it raineth, for loytering men." *Five Hundred Points of Good Husbandry* (1638), p. 85.

62. Overton, *Agricultural Revolution in England*, p. 41.

63. Bartholomew Dowe, *A dairie Booke for good huswiues* (London: Thomas Hacket, 1588), A3v.

64. Markham, *English Hovse-Wife*, p. 194. In families where agriculture centered on sheep rather than cows, the wife might be involved in spinning, dyeing, and carding wool as part of her daily chores.

agricultural work, another daughter, Mary, got "my customary Storehowse & the wharfe being pcell of the old Docke wth th'appurtencs neere the water side in Wivenhoe," and the will mentions four boats: "my Catch or crayer called the grigory," "my litle Catch or Boat called the litle John," "my Katch or crayer called the great John," and "the shipp called the Mary Ann."⁶⁵ This testator's engagement in a dual economy is clear to see. The bequests of a tanner with animals seem small in comparison, but they still reflect a substantial herd. William Fuller (not Christopher Fuller's son) bequeathed to his wife Judeth "ten pound and five ~~x~~ beasste [cows] ~~a~~ and one wennelll and ~~five~~ foure sheepe," to his sone Joseph "all my ~~leather~~ leather and barke and all tubes and tules belonging therevnto," and over £80 more in monetary bequests.⁶⁶ While the animals' skins might be destined to become leather in his Woodham Walter tannery, the "beasste" were first used for milking.⁶⁷

Despite the fact that Anthony Lord of Hatfield Broad Oak referred to himself as a "mercer and husbandman" in his 1625 will (the only testator to list himself as having a dual occupation), there is nothing in it that hints at his position as mercer (a dealer in textiles). Rather, the will included two "Tenemente and pcells of Land wth thapp'ten'cs" to his mother Joane Lord, who was forgiven the 28s she owed him. In addition, Lord left money totaling £8 to his niece and two nephews; and "the rest & residue of my goods Leases Corne Cattell Corne on the ground sheepe & Lambes & all other debtes chattles goods & ymplements of husbandry whatsoever by this my will vnbequeathed" to his brother Richard, whom he made his executor.⁶⁸ Lord's will opens up the possibility that numerous other testators who termed themselves husbandman might have had second occupations that are not made visible by their possessions.

Almost all male occupational/status groups with thirty or more testators in the ERO dataset, including individuals who listed themselves as laborers, show a decline in proportion in the group of 451 wills that include specific animal bequests.⁶⁹ The exceptions are yeomen and husbandmen. The

---

65. ERO, D/ACW 8/261 (1620), made probate 56 days after writing. A crayer is a "small trading vessel"; *Oxford English Dictionary*, s.v. "crayer."

66. ERO, D/ABW 48/173 (1626), made probate 15 days after writing.

67. As noted in the introduction, the inventory that followed the will of Richard Cesar of Terling included "his shoppe tooles" as well as a reference to bullocks and a little hog. ERO, D/ACW 9/62 (1622), made probate 66 days after speaking.

68. ERO, D/ABW 46/14 (1625), made probate 13 days after writing.

69. The figures for these groups are gentlemen, 96 wills (9 animal bequests); carpenters, 88 wills (5 animal bequests); tailors, 85 wills (2 animal bequests); weavers, 73 wills (3 animal bequests); laborers, 60 wills (5 animal bequests); blacksmiths, 39 wills (5 animal bequests); clothiers, 39 wills (2 animal bequests); butchers, 34 wills (2 animal bequests); tanners, 33 wills (2 animal bequests); millers, 31 wills (2 animal bequests); shoemakers, 30 wills (0 animal bequests).

proportion of yeomen changes notably: while they wrote 25 percent of wills in the large ERO dataset, their testaments make up 36 percent of those that include animals. With husbandmen, the difference is not so marked, but it is still present: they wrote 16.5 percent of wills in the large ERO dataset but 18 percent of those that include animals.[70] In addition, 14.5 percent of all yeomen bequeathed animals, as did over 11 percent of husbandmen. Excluding these two groups, the average percentage of wills by all other male testators that include animals is just under 8 percent, although the proportion of laborers leaving quick cattle is slightly higher. Overall, then, yeomen and husbandmen, the groups most closely associated with agricultural production who were likely to own their own livestock, were also more likely than their neighbors to bequeath animals specifically. Of the 473 wills of individuals in the ten most represented trade groups in the large ERO dataset—carpenters, tailors, weavers, blacksmiths, butchers, clothiers, tanners, millers, shoemakers, and glovers—only 25 included quick cattle. Thus, although individuals from these trades account for more than 10.5 percent of wills overall, they make up just over 5 percent of wills that include livestock.

This does not, of course, tell us that these tradesmen did not own quick cattle. What it does tell us is that they were less likely to specify them in their wills than yeomen or husbandmen. This could signal, as I proposed in the previous chapter, that they took care by avoiding specificity and relied on general terms to distribute their livestock. However, that would suggest that yeomen and husbandmen, groups who were more likely to bequeath specific animals than tradesmen, were less careful on those terms. I think that what we can see, in fact, are two manifestations of care in the methods chosen by testators: one that was intent on conveying quick cattle to human kin by using generalization and another that recognized animals as both valuable stock and individual creatures. The two are not in contradiction, I suspect: they are simply different means of distributing possessions. But they do reveal different perceptions of those self-moving movable creatures.

---

70. Two other groups show slight increases in proportion of those who included specific animal bequests. Those who termed themselves "singlemen" wrote sixty-three wills in the large ERO dataset (1.5 percent of the total). Of this group, eight wrote wills that included animals, just under 2 percent of those specifically bequeathing animals in the large dataset. Because of the lack of clarity as to what "singleman" might mean (it could be equivalent to "widow" and so might indicate marital status rather than occupation), it is possible that some of those listed as singlemen might be, for example, husbandmen. The other group for whom the proportion increases in relation to animal bequests are male testators with no occupation listed. This group makes up 12 percent of wills in the large ERO dataset and 14 percent of wills with specific animal bequests. Once again, many of these testators might be yeomen or husbandmen.

The proportion of women who bequeathed quick cattle is also worth noting. That some of them were engaged in agricultural production is obvious, but it is more difficult to say how many because their testamentary occupation/status labels were (usually) devoid of clarity—only one woman listed herself as the "wife of a husbandman," and she did not bequeath any animals, probably because they were legally owned by her spouse.[71] The other occupation/status labels (and numbers) for female testators are widow (736—one of whom lists herself as "innholder and widow," signaling that other women may have been engaged in trade[72]), singlewoman (58), spinster (28), wife (10), maid (2), daughter (1), gentlewoman (1), and servant (1). In addition, forty-seven wills of women do not include an occupation/status label. Female testators make up 19 percent of the large ERO dataset but just under 16 percent of all testators who included specific animal bequests.[73] And while the proportion of women who included such bequests in their wills was the same as that of male testators who are not yeomen or husbandmen, overall, 8 percent of women included bequests of animals in their wills, in contrast to just under 11 percent of all men.

When the focus is put on the bequeathing of particular species, however, something else emerges, and the wills reflect the division of labor Markham outlined in a way that is especially marked in relation to two sets of animals: poultry and horses. While testaments by women make up just under 16 percent of wills that specifically bequeathed animals, 32 percent of the (albeit small sample of) twenty-two wills that include bequests of birds (poultry, hens, geese, ducks) are by women, while only 9 percent of the 143 wills that bequeath horses are by female testators. Perhaps even more telling is the fact that 77 percent of the recipients of bequests of birds were women.[74] By comparison, only 40 percent of wills that bequeath horses left them to female legatees.[75] The proportion of women who bequeathed cows and pigs is closer to the overall average of women testators who included specific animal bequests: over 16 percent of wills that include cows are by women, as are

---

71. ERO, D/ACW 12/33 (1634), made probate 76 days after writing.

72. ERO, D/ABW 49/166 (1629), made probate 109 days after writing.

73. Widows on their own make up 17 percent of all wills in the large ERO dataset, but just under 14 percent of wills that included specific animal bequests.

74. Just over 18 percent of testators left birds to both male and female legatees; only one (4.5 percent) left poultry to a male legatee alone.

75. Just over 10.5 percent left horses to both male and female legatees; 53 percent left them to male legatees alone.

17 percent of those that include swine. But, once again, the identity of the legatees signals something else. In the 224 wills that bequeath them, 320 legatees received dairy cows, over 58 percent of whom were female.[76] In wills that include swine, that figure is 68.5 percent.[77] The difference in the percentage of female recipients for different animals—the fact that female legatees received (in decreasing proportion) birds, pigs, cows, and horses—might reflect the animals' financial value. Even as cows and pigs were, like poultry, associated with female labor, as the animals increased in worth they were more likely to be bequeathed to male legatees.

The distribution of animals in Christopher Fuller's will reflects in some ways what the larger body of wills also reveal. Christopher, the eldest son, got the team of horses and the cart and plow, three cows, a heifer and a calf, a sow, two shots, and some piglets; William got one old cow, two steers, a heifer and two calves, a sow, two shots, and some piglets; while Alice got just one cow, one heifer, and one calf. However, what Alice lacked in legacies from outside the house, she made up for in those from within. Apart from the beds and bedding, which were shared fairly evenly (although Alice got the new featherbed, the best flockbed, the best coverlet, and two pillowcases), Fuller gave almost all of his furniture to his daughter. Christopher got the smallest chest, William got nothing, and Alice got everything else—a table, a stool, chairs, a chest, and a cupboard. The housewife got the household stuff.[78]

"Alce Milles" as well as being a relation, may also have been a maid in the house and dairy, with her legacy of a bed and bedding serving as recognition of this. However, while we cannot know her role in the household (or even if she had one at all) from the will, the differing ways Fuller describes his animals can tell us a great deal about how they were organized on his property and in his mind. By reading his will alongside others in the large ERO dataset and in the context of contemporary agricultural texts, we can see how typical life on Merits was at the time it was written; and we can glimpse something more about livestock and, by default, about the people who lived and worked with them as well.

---

76. Just under 15.5 percent left animals to both male and female legatees; 28 percent legatees left animals to males alone. Although 224 wills in the animal dataset included cows, five (2 percent) specified that the cows should be sold and so did not bequeath them. These five wills are not included in the figures for legatees here.

77. Nine percent left the animals to both male and female legatees; 22 percent left them to males alone. In the figures for all of these animals, I am not including the general bequests of animals in the collecting phrases of wills; I am only including specific bequests.

78. Christine Churches outlines an alternative reading of references to property in wills from a slightly later period than I discuss here in "Women and Property in Early Modern England: A Case-Study," *Social History* 23, no. 2 (1998): 165–80.

To begin where Christopher Fuller started his survey of his animals: he did not individualize his horses, he simply referred to them as "my teame."[79] This designation, along with the fact that he did not split up the "teame" into different bequests (some to one child and some to another) tells us that it was the horses' role as plow and cart animals that was important to him. This, in turn, reveals that he did not own a horse just for riding, which reinforces my reading of him as a self-sufficient but not large-scale yeoman. Such grouping of horses in wills was not always the case: they were frequently represented individually.[80] Indeed, over 46 percent of wills in the large ERO dataset that include bequests of horses give descriptions of an animal that would allow it to be individualized: examples include "the Ronded nagge yt Mr Maxye did ryde vppon,"[81] "my bay nagg with a white foote,"[82] and "a Mare Colt with a bald fface."[83] Five horses were also given names—an issue I return to in the following chapter.

The wills also show that different horses had different functions. Ambling horses, smooth-paced animals that were good for riding, appear in a number of wills.[84] William Greene of Hadstock bequeathed to his son Bennett a horse that would have been invaluable in his job as park keeper: "his stalkinge horse."[85] And when William Porter of Tendring bequeathed to his wife "my dun horse, my gray horse, & my forehorse" he may not simply have been individualizing them for clarity, he may also have been recognizing their distinct roles and perhaps personalities.[86] The fore-horse was the leader of the

---

79. West Ham miller, John Grubb, also bequeathed his horses as a group, leaving "my large carte and the horses and furniture to them belonging" to his son. ERO, D/AEW 18/155 (1627), made probate 60 days after writing.

80. In the large ERO dataset, just under 60 percent of the 77 wills that include a particular bequest of only one horse mentioned other animals. For the 66 wills that included more than one horse, 70 percent also contained other animals. This might be early evidence of horses being used for riding purposes by individuals not involved in agriculture.

81. ERO, D/ABW 46/218 (1624), made probate 30 days after writing. I have been unable to discover what "ronded" meant.

82. ERO, D/ABW 43/44 (1621), made probate 26 days after writing.

83. ERO, D/ABW 47/15 (1626), made probate 113 days after writing.

84. For example, D/ABW 43/71 (1621), made probate 25 days after writing; ERO, D/ACW 9/229 (1625), made probate 36 days after writing; ERO, D/ACW 10/85 (1625), made probate 200 days after writing; ERO, D/ACW 10/66 (1626), made probate 18 days after writing; ERO, D/ABW 49/150 (1629), made probate 62 days after writing; and ERO, D/AEW 19/243 (1633), made probate 73 days after writing.

85. ERO, D/ACW 11/255 (1632), made probate within 18 days of speaking. The *OED* defines a stalking horse as "a horse trained to allow a fowler to conceal himself behind it or under its coverings in order to get within easy range of the game without alarming it."

86. ERO, D/ABW 47/195 (1625), made probate 90 days after writing.

team, and the limitations of this vital role were laid out in Nicholas Breton's 1618 dialogue between a Countryman and a Courtier. In opposition to what he perceived as the attractive artifice that can come from urbane living, the Courtier likened his cousin's dull naturalness to one "bred like a fore-horse to goe alwayes right on, and rather draw in a cart, then trot in a better com-passe."[87] Thus, when Stephen Kingstone, a blacksmith from Thorrington, dif-ferentiated among the horses he bequeathed—leaving "my forehorse & my nagge" to his wife Elizabeth, "my horse called Bucke" to one son, and the "piebald nagge" to another son—he was reflecting on the ability of some but not all to "trot in a better compasse."[88] The named horse was, perhaps, the favorite animal that he had ridden during his life. Fuller's choice of words to describe his "teame of horses" does not discount the distinctiveness of each of the animals: as Markham's outline of the husbandman's day shows, even animals perceived as a group had to be tended individually. What Fuller's description of them in his will shows, though, is that their individuality was secondary to their role as part of a collective.

Fuller also mentioned numerous swine in his will: two sows, four shots (young pigs), and two litters of piglets. We might divide this into two groups, each containing one sow that had two shots and one new litter (this is how they are bequeathed—one porcine family to each of his sons). Such numbers seem to offer support for some of the printed statements from the period about the reproductive capacity of sows. At the end of the sixteenth cen-tury, for example, Leonard Mascall wrote that sows might "haue pigges twise, some thrice a yeare,"[89] an estimation that Gervase Markham echoed: "A Sow will bring forth Pigs three times in a yeere, namely at the end of euery ten weekes, and the numbers are great which they bring forth: for I haue knowne one Sow haue twenty Pigs at one litter." His estimate of the gestation period is less than Mascall's (who says that it is four months long) and may exag-gerate the animals' productivity, but Markham was firmly guided by reality when he recognized that "a Sow can bring vp no more Pigs than she hath Teats."[90] The number of teats ranges by breed (twelve is a modern average), and writing in the mid-sixteenth century Thomas Tusser proposed that "Of one sow together, reare few above five, / and those of the fairest, and likeliest

87. Nicholas Breton, *The Court and Country, Or A Brief Discourse Dialogue-Wise Set Down Between a Courtier and a Countryman* (London: G. E. for John Wright, 1618), B3v.

88. ERO, D/ABW 51/171 (1633), made probate 1,289 days after writing.

89. Mascall, *Government of Cattle*, p. 256.

90. Markham, *Cheape and Good Hvsbandry*, p. 124; and Mascall, *Gouernment of Cattle*, p. 256.

to thrive."[91] The two wills in the large ERO dataset that include sows and specific quantities of piglets give different numbers per sow. The widow Helster Woulfe of Rickling left her daughter "my greatest sowe & six piggs,"[92] and Thomas Huskyn, a carpenter from West Ham, left "a sowe and Itm Tenne Piggs" to be shared by his son and daughter.[93] Woulfe's will was made probate fifty-six days after it was written and Huskyn's after sixty-five days, which signals that it is likely that each testator died within about a month of writing their will (Fuller's will was written within a week of his burial in June 1620, and was made probate after seventy-one days).[94] This allows us to assume that testators were writing about existing animals rather than bequeathing future births—and thus were not echoing the bequest of a "Ewe lambe" four years after the testator's death mentioned in the previous chapter. For this reason, the dates these wills were written might tell us something about the breeding of swine in the early seventeenth century. Woulfe's will was written on 23 March, Huskyn's on 9 July, and Fuller's on 22 June. The three litters per year that Mascall and Markham mention fit: the wills reveal a late winter/early spring litter (hence Woulfe's piglets in March), a late spring/early summer one (Fuller's and Huskyn's litters in June/July), and perhaps one more in the autumn. While neither Mascall nor Markham mention an age for weaning piglets, we can assume that the appearance of a new litter would require this. But Markham does give a sense of how long an animal might be referred to as a shot: these, he says, "are Swine of three quarters, or but one yeere old [and] are the daintiest Porke."[95] Given the possible age range of shots, the existence of them in wills written in February (two wills),[96] June (Fuller's will) and November,[97] does not offer insight into the detail of the month of their likely birth, but it does show that they are likely to have been present on farms throughout the year.[98]

Fuller's inclusion of swine of different ages also reveals the reality of pork production in the period. Many piglets might be born, but not all could be fed by their mother, nor could all those that had been fed be supported by

---

91. Tusser, *Five Hundred Points of Good Husbandry*, p. 64.

92. ERO, D/ABW 48/52 (1627), made probate 56 days after writing.

93. ERO, D/AEW 17/299 (1625), made probate 65 days after writing.

94. No parish register from the period exists for Rickling or West Ham, so we cannot date the burials of Woulfe or Huskyn.

95. Markham, *Cheape and Good Hvsbandry*, p. 125.

96. ERO, D/ABW 48/112 (1627), made probate 33 days after writing; and ERO, D/ABW 49/229 (1628), made probate 98 days after writing.

97. ERO, D/ABW 44/187 (1622), made probate after 7 days.

98. One will written in April also included shots (ERO, D/ACW 11/106 [1630]) but this was made probate 786 days after writing. I discuss this will below.

the food that was available in the yard after weaning. For this reason only a few might be allowed to survive infancy to grow bigger; the others dying by disease or culled. In Fuller's household, it is possible that two shots survived from each sow's previous litter. The unwanted piglets could have been consumed by the household but were, more likely, sold on: "sucking pig" was a recognized delicacy. Mascall, again, offers insight here: for those who live "in places and villages nigh vnto great townes, or Gentlemens houses," he says, "inselling the yong sucking pigs" is common, and such pigs are "alwaies ready money to" their owners.[99] Indeed, Tusser proposed in his outline of "Ianuaries Husbandry" that "Who eateth his veale, pig, and lambe . . . / shall twise in a weeke, goe to bed without broth."[100] Consuming your own young livestock was poor financial planning.

The piglets that survived continued to grow to be fattened for winter slaughter, and this, for Tusser, was the best use of these animals: "he that can reare of a Pig in his house, / hath cheaper his bacon, and sweeter his souse."[101] Ralph Josselin, the vicar of Earls Colne, seemed to follow this advice. In his diary entry for 19 December 1653, he noted that he had "kild my hogge he proved very cleane and most excellent meate," and again, on 19 November 1660, he wrote that "we kild a good hogge, which proved neat and cleane, a mercy to bee observed."[102] Nicholas Breton presents another kind of value in swine in his 1618 dialogue: "when wee kill Hogs [we] send our Children to our neighbours with these messages," the Countryman states: "My Father and my Mother have sent you a Pudding and a Chine, and desires you when you kill your hogges you will send him as good againe."[103] There is a practical logic to this generosity: fresh meat goes off quickly, so sharing your animal's carcass with a neighbor on the understanding that the neighbor will later share theirs with you ensured that fresh meat would be available over a longer period than keeping the whole animal to yourself would allow. Undoubtedly, Breton's text offers an idealized vision of country life, but it is interesting that

---

99. Mascall, *Government of Cattle*, p. 256.

100. Tusser, *Five Hundred Points of Good Husbandry* (1638), p. 60.

101. Tusser, *Five Hundred Points of Good Husbandry* (1638), p. 60. Two wills refer to "store pigs," or young animals kept for breeding or fattening (the latter usage, according to the *OED*, first appeared in 1620). Mary Pannell of Shalford bequeathed to her son "one stoare pigge now in the yard" and William Smieth of Great Sampford bequeathed "twoe stores." ERO, D/ABW 46/74 (1625), made probate 72 days after writing; and ERO, D/ABW 44/315 (1623), made probate 37 days after writing.

102. Ralph Josselin, *The Diary of Ralph Josselin 1616–1683*, edited by Alan Macfarlane (Oxford: Oxford University Press, 1976), pp. 315, 472.

103. Breton, *Court and Country*, D1v–D2r. The "pudding" is "the stomach or entrails"; and the chine is "the backbone and adjoining flesh"; *Oxford English Dictionary*, s.v. "pudding"; s.v. "chine, n.2."

he places swine—what Markham calls "sinkes" and consumers of "ordure"—at the heart of his conception of community.[104]

Wills most commonly use the term "hog" when referring to porcine animals: it appears in thirty-one of the fifty-four wills in the large ERO dataset that include swine (the term "swine" on its own was used in only two wills).[105] A hog, according to the *OED*, is "a domestic pig reared for slaughter; *spec.* a castrated male pig. Also more widely: any domestic pig."[106] The term's wide meaning—interchangeable with "swine" in one will[107]—would seem to offer little help with specifying the sex of the animals being bequeathed.[108] Some do specify sex, though, signaling that not all "hogs" were male. David Tarner of Great Oakley, for example, bequeathed "one sowe hodge,"[109] and Robert Freeman left "one gelt ~~one~~ sowe hogge."[110] "Gelt" here would mean "gelded," and Mascall offers directions on how to perform this operation.[111] "Sow" as a term on its own appears in just five wills (including Fuller's):[112] the will of Nathaniel Butler of Bowers Gifford listed a "great sowe,"[113] and that of

---

104. Markham, *Cheape and Good Hvsbandry*, p. 123.

105. ERO, D/ABW 47/121 (1626), made probate 59 days after writing; and ERO, D/AMW 1/110 (1624), made probate 53 days after writing.

106. *Oxford English Dictionary*, s.v. "hog."

107. ERO, D/AEW 19/255 (1634), made probate 131 days after writing: "all my hoggs or swine."

108. Peter Fish of Tollesbury bequeathed "one hogg" (ERO, D/ABW 43/115 [1620], made probate 27 days after writing); John Collins of Shopland left "two of my best Hoggs" (ERO, D/ABW 50/151 [1631], made probate 78 days after writing); and Richard Cesar, Widow Sysaye's husband, left "one little hogge," a phrase that creates some ambiguity as to the age of an animal referred to by this term (ERO, D/ACW 9/62 [1623], made probate 66 days after speaking). Confusingly, Markham notes that "hog" is a term that can also be used to refer to lamb in its first year, but he specifies "wether hog" (male) and "ewe hog" (female). Of the wills in the animal dataset that refer to hogs, some use that term on its own and so could plausibly refer to sheep, however, many use the term with an adjective—barrow hog, sow hog, etc.—that makes clear that these are swine, as does the interchangeable use of the term with "swine" in ERO, D/AEW 19/255. Because no use of "wether hog" or "ewe hog" exists in the dataset, I have taken all references to hogs to mean swine. Markham, *Cheape and Good Hvsbandry*, p. 107.

109. ERO, D/ACW 9/232 (1625), made probate 32 days after writing. This term is also used in ERO, D/ACW 9/76 (1623), made probate 28 days after writing.

110. ERO, D/ACW 9/37 (1623), made probate 986 days after writing.

111. "Some thinke in spaying them of shootes is best, cutting them in the mid flancke with a sharp knife two fingers broad, in taking out the bag of birth, and cutting it off, and so they doe stitch vp the wound againe, and then annoint it, and keepe her warme in the stye two or three dayes after." This risky operation was done for a simple reason: "they will ware the fatter" because of it. Mascall believed, however, that more money could be made from a productive sow than a fat one. Mascall, *Gouernment of Cattle*, pp. 256–57. The *OED* records "gilt" as also meaning "a young sow or female pig," but the inclusion of both "gelt" and "sowe" in Freeman's will seems to imply spaying.

112. ERO, D/ACW 11/106; ERO, D/ABW 43/92 (1620), made probate 57 days after writing; ERO, D/ACW 8/270; ERO, D/ABW 43/168 (1621), made probate 95 days after writing; and ERO, D/ABW 46/107 (1625), made probate 494 days after writing.

113. ERO, D/ABW 49/66 (1630), made probate 45 days after writing.

Humphrey Cook of Bulphan mentioned halfe the sheates and the litle sowes," which seems to show that he used "shot" to refer only to young male animals, thus revealing the possibility that Fuller kept only male swine to fatten.[114] As the shot matured it could be given another name: a "barrow [castrated] hog" is mentioned in the wills of two testators[115] and a "great barrowe pigge" in another, The latter usage signaling that "pig" obviously did not just mean piglet in this period.[116]

All these descriptions of swine not only allow for easier distribution of legacies by executors, they also give insight into life with hogs in early modern Essex. However, out of 4,444 wills in the large ERO dataset, only fifty-five (1 percent) include specific mention of swine (by comparison, 271 wills—6 percent—include bequests of bovine animals). One percent of wills does not seem to be many for an animal that was often viewed as inexpensive to keep, and suggestions as to why swine are so rarely referred to can be found in the work of Francis W. Steer. He proposed that their relative absence in inventories from Essex from 1635 to 1749 could be for two reasons; first, because in that county, "a great area of the woods had been cleared with an attendant loss of natural food supplies" for pigs. Second, because "it was customary to kill pigs, other than those reserved for breeding, after the acorns and beech-mast had been consumed and before the hard winter days set in."[117] The wills in the large ERO dataset, which slightly predate Steer's sample, seem not to support his second point and suggest, perhaps, that a change in husbandry took place over the course of the seventeenth century. Of the fifty-five wills that include swine, thirty (55.5 percent) were written between the beginning of November and the end of March, and more wills written in February (eleven) mention swine than those written in any other month of the year (none appear in a will written in August).

---

114. ERO, D/ABW 48/235 (1626), made probate 19 days after writing.

115. ERO, D/ABW 43/92 (1620), made probate 57 days after writing; ERO, D/AEW 19/299 (1634), made probate 50 days after writing.

116. ERO, D/AMW 1/164 (1632), made probate 139 days after writing. Pigs are referred to in eighteen wills. One will (ERO, D/ABW 50/146 [1631], made probate 75 days after writing) referred simply to "pigges," but another (ERO, D/AEW 17/323 [1625], made probate 82 days after writing) referred to "my six piges"; another (ERO, D/ABW 51/56 [1632], made probate 23 days after writing) mentioned "two piggs"; and yet another (ERO, D/ACW 8/261 [1620], made probate 56 days after writing) bequeathed "two of my best pigges."

117. Francis W. Steer, "Introduction," in *Farm and Cottage Inventories of Mid-Essex, 1635–1749*, edited by Francis W. Steer (Colchester: Wiles and Son, 1950), p. 56. On the decline of pigs and the loss of woodland, see also Richard Thomas, "Of Books and Bones: The Integration of Historical and Zooarchaeological Evidence in the Study of Medieval Animal Husbandry," in *Integrating Zooarchaeology*, edited by Mark Maltby (Oxford: Oxbow Books, 2006), p. 21.

There are two possible interpretations of the persistence of swine in wills from Essex in the period of the large ERO dataset. First, if Steer's findings for the period after 1635 were replicated in the earlier period, then the absence of swine from inventories and their continuous inclusion in wills might show us that sometimes testators made wills before death was imminent and included bequests of animals that had not yet been born, thus revealing that wills did not only distribute what already existed, but also planned for a fruitful future. The husbandman Christopher Homan, for example, bequeathed "all my Cattell sixe Cowes and two sheepe (one sowe &) to shoates" to his wife and two sons, Christopher and William; "two lames" to a third son, John; and "a mare Coltte [and] all my foueles" to Christopher in his will, which was written on 23 April 1628. This was just over two years before his burial in the churchyard of St Mary, Wix, on 9 May 1630 (the will was made probate just over a month after his burial).[118] It is worth considering how to interpret a will's inclusion of specific animals when it was apparently written so far in advance and when the "decay" of animals would have been something any testator must have taken into account—especially with pigs, as they were often short-term residents, born for slaughter rather than breeding. It is possible that Homan firmly believed at the time he wrote his will that he was about to die and that it reflected his possessions as they were in April 1628 and that he then recovered but did not alter or replace his will as time wore on as it continued to reflect accurately—or accurately enough—his testamentary wishes and his possessions (in which case shots must have been a constant presence in his yard). Or it might be that he wrote the will in advance, as clergy urged people to do—"being sicke in body," after all, was simply part of the human condition. Perhaps Homan felt he could do this and be specific in how he bequeathed his animals because he was sure that he would continue to raise livestock and would always have the animals he wished to leave his family (again, this would imply shots as almost permanent presences). This is plausible but would place the executor—in this case his son—in a potentially difficult position because if there was a shortfall between the actual possessions and the bequests of a testator, the law would require the executor to make it up.

We should, perhaps, regard Homan's writing of his will so far in advance of his death as unusual: only 15 percent of testators who included specific animal bequests wrote wills more than a year before the probate date; in the sample dataset, that figure is 19 percent. On balance, it seems to be the case that most

---

118. ERO, D/ACW 11/106 (1630), made probate 786 days after writing; Register of Baptisms, Marriages and Burials, St Mary, Wix, ERO, D/P 172/1/1 (1560–1675).

testators wrote a will close to death. And it is notable that of the wills that included bequests of swine, over 40 percent were made probate within sixty days of being written (in the sample it was 38 percent) and 77 percent were made probate within 180 days (64 percent in the sample). Indeed, of the wills that include swine that were written in the months that Steer found them to be absent from inventories (November to March), 45 percent were made probate within two months, over 80 percent were made within six months, and only 16 percent were made probate over a year after the will was written.

Because it is likely that most wills were written close enough to death to have a realistic sense of animal property, pigs may have been included in wills only rarely for another reason, and the Great Braxted widow Anne Birch's unhelpfully vague bequest of "^halfe a load of hoggs^," inserted as an after-thought into an already written line in her will, might give a clue.[119] Perhaps the fact that pigs did not require or receive the same kind of attention as horses, or the same close contact as cows, coupled with their lower value and their temporariness in the household (the fact that most were bred for slaughter rather than for long-term productivity) might mean that they were frequently included in the collecting phrase at the end of the writing of a will. And pigs were even more often omitted by those who left nuncupative wills than by those who had prepared formally written documents. In the large ERO dataset, 14 percent of wills are nuncupative; that proportion decreases to 11 percent in the group of 451 wills that include specific animal bequests. But the drop in proportion was even more significant when it came to the bequeathing of swine: of the fifty-five wills that included swine only 5.5 percent of them were nuncupative (and no will that left poultry was nuncupative.)[120] This suggests that if pigs were constant possessions—and in Stanton-by-Bridge in Derbyshire between 1537 and 1755, D. J. Baker found that "nearly everyone kept one or two pigs"—then in their spoken deathbed testaments, testators more frequently overlooked their swine than they did their sheep, cows, or horses.[121]

If pigs, more than any other mammalian livestock animal, were left out of wills, when they were included, they went without individualization too—and the two things might be linked. Testators used only age, sex, and

<hr/>

119. ERO, D/ACW 11/222 (1631), made probate 11 days after writing.

120. Over 13 percent of wills bequeathing sheep were nuncupative, whereas of wills bequeathing horses, only just over 8 percent were nuncupative, and of wills bequeathing cows, only 7 percent were nuncupative.

121. Baker, "Stanton-by-Bridge," p. 83.

genital intactness (with a brief reference to size in Nathaniel Butler's "great sowe" and Humphrey Cook's "litle sowes") to describe them. Even sheep—animals that were kept in flocks potentially at a distance from the home and were herded by a communal shepherd[122] and perhaps only came in for close attention at particular times of year such as lambing and shearing—are described individually on two occasions in the large ERO dataset: John Scarf of Little Hallingbury, a laborer, left his daughter Joane "my brittle face sheep being pold"[123] and in a memorandum at the bottom of his 1623 will, Launcelot Brytten, a butcher who lived in Wethersfield left his "honest & kynd neighbor" Germany Cranfford "a horned wether wth two teeth."[124] In contrast, testators frequently used clear and individualizing language to demarcate bovids. While only ten (4 percent) of the 217 wills that bequeath sheep include descriptions that go beyond age or sex,[125] the proportion of wills that use specific terms to describe bovine animals is 35 percent (the figure is 46 percent for horses). This detail is reflected in Fuller's will, where he defines bovids by age and sex, as was conventional, but also by size and breed or color—and one by name. These methods, which are present in a third of wills that include bovids, not only helped executors carry out the testators' last wishes, they can also tell us much about life lived with cows in early modern Essex.

Where the collecting phrases that were the focus of the previous chapter might be read as evidence of care for human recipients in the bequeathing

---

122. On collective shepherding in the New World, see Susan M. Ouellette, "Divine Providence and Collective Endeavor: Sheep Production in Early Massachusetts," *New England Quarterly* 69, no. 3 (1996): 355–80. In a dispute over the enclosing of common land in Newport, Essex, that came before the Star Chamber in 1620, one deponent refers to "the comon heardman of the said towne." National Archive (hereafter NA), "Attorney General v Nightingale," STAC 8/34/5 (1620). Two testators listed themselves as shepherds in the large ERO database: see ERO, D/ABW 46/110 (1625), made probate 645 days after writing; and D/ACW 11/46 (1629), made probate 19 days after writing. Neither included any animal bequests in their will.

123. ERO, D/ABW 46/33 (1625), made probate 548 days after writing. "Brittle" here possibly means fragile or weak and "pold" means hornless. *Oxford English Dictionary*, s.v. "brittle."

124. ERO, D/AMW 2/50 (1623), made probate 88 days after writing. It is unclear whether the marking referred to in the nuncupative will of Ralph Bynder of Cold Norton refers to one or all of his animals. He bequeathed "vnto his Cousyn John Bynder the School Mr Tenne Ewe sheepe, & a Rame, remayneinge wth Thomas Humfrey, of Purleighe laborer, beinge branded wth a brande like a Belle." ERO, D/ABW 46/174 (1625), made probate 9 days after speaking.

125. These include the two descriptions given above, plus five wills that include black ewes and/or lambs (ERO, D/ABW 43/71 [1621], made probate 25 days after writing; ERO, D/ABW 48/113 [1627], made probate 46 days after writing; ERO, D/ABW 52/16 [1633], made probate 80 days after writing; ERO, D/ACW 11/78 [1629], made probate 16 days after writing; ERO, D/AEW 18/76 [1626], made probate 1,579 days after writing), two that include russet ewes and lambs (ERO, D/AEW 19/111 [1631], made probate 138 days after writing; and ERO, D/AEW 19/299), and the will with the branded sheep. I return to lambs in chapter 4.

of animals (the general terminology allowing room for changes in livestock population), it is possible that descriptive specificity might offer evidence of another kind of care: this time for particular animals. Pigs were often short-term residents, sheep were largely anonymous flocks, and horses could be members of teams, but cows in particular (I come to bulls, bullocks, weanels, and calves later) had a special place on the early modern farm. Not only were they valuable, they were also incredibly important because of the nutrition they provided. In addition, they were physically close to members of the household: someone—perhaps the wife, daughter, or maid—leaned up against a cow's flank twice a day, day-in, day-out. These animals mattered—in every possible way.

Beginning with his emphasis on the color of his cows, Fuller bequeathed two "brended" animals. The *OED* defines "brended" as "of a tawny or brownish colour, marked with bars or streaks of a different hue; also *gen.* streaked, spotted; brindled."[126] Other wills in the large ERO dataset include references to such streaked cattle. For example, John Archer of Bradfield, who was concerned about the possible "decay" of his cows, left "unto Michaell Archer my sonne one Cowe called the brinded cowe."[127] Husbandman Richard Taylor's bequest to his son Richard of "the branded Cowe" probably reflects variant early modern spelling rather than indicating that he had used a hot iron to mark his animal, because it is likely that if he had branded one, he would have branded them all and the description would not have helped specify an individual.[128]

Fuller had two brended cows: one described as "my old great brended Cowe" and one that he described simply as "my brended Cowe." The age of the former perhaps made her less valuable than the latter because she had less productive time left, but, alternatively, she may also have been a tried and tested breeder and may have been more highly esteemed than other cows (the fact that Fuller bequeathed her to William, the second son, however, might signal that her value was lower than that of other cows). Fuller's will thus challenges the idea that all

126. *Oxford English Dictionary,* s.v. "brended."

127. ERO, D / ABW 47 / 212 (1626), made probate 78 days after writing. See also ERO, D / AEW 18 / 210 (1628), made probate 71 days after writing; and ERO, D / ABW 49 / 229 (1629), made probate 98 days after writing. Other wills include references to "flecked" animals, which may have been of a similar appearance to the brended ones. See, for example, ERO, D / ACW 9 / 226 (1625), made probate 97 days after writing; ERO, D / ABW 44 / 14 (1622), made probate 38 days after writing; and ERO, D / AEW 17 / 188 (1624), made probate 113 days after writing.

128. ERO, D / AEW 19 / 76 (1631), made probate 144 days after writing. See also ERO D / ABW 50 / 82 (1631), made probate 16 days after writing; and D / AEW 19 / 233 (1633), made probate 269 days after writing.

old animals would be sold on (as Mary Archer proposed in her will), as does the will of Roger Hull, a miller from Great Maplestead. Among Hull's numerous animals is "my ould Cowe," which he bequeathed to his daughter Margaret.[129] As in the case of Fuller's aged beast, it is worth pausing over the status of this elderly creature. If she had reached bovine menopause after a long and productive life her function seems unclear. She may have been kept as a companion for other cows—a calm animal who reduced stress levels in Hull's herd. Or perhaps she was kept as an animal to pull the plow, as Heresbach noted. Or maybe Hull was simply attached to her. Such attachment to ageing animals is impossible to prove beyond the glimpses that we get of the devastation the death of a cow might cause. But that devastation, of course, does not have to be about the animal's death: it could be about the impact on the humans: "No cow" could be a cry from the belly or the pocket as much as from the heart.

But perhaps use of the term "old" in wills is comparative: Fuller's animal is old only in contrast to the other brended cow in his will. But then "eldest" would be the appropriate term—and Fuller used that in relation to his sons. "Old" does seem to imply an animal beyond the age of utility, and thus the suggestion that a close, long-term relationship between testator and animal might be reflected in the will should not be automatically dismissed. Wills are about care for present and future generations, and it is possible that such care extended to the well-being of animals, particularly to animals with whom the testator had a long-term relationship.

Color was another way to distinguish cows in wills. Fuller differentiated his brended cows from the two black northern cows, who were, in turn, differentiated from each other not by age but by size: one is "litle" and the other "great." Markham described "northern" cows as "bred in *Yorke-shire, Darby-shire, Lanca-shire,* and *Stafford-shire*" and as "generally all blacke of colour," noting that "they whose blacknesse is purest, and their haires like Veluet, are esteemed best." He continued: "they haue exceeding large hornes, and very white, with blacke tippes; they are of stately shape, bigge, round, and well buckled together in euery member, short ioynted, and most comely to the eye, so that they are esteemed excellent in the market."[130] Joan Thirsk has described them as "good for tallow, hide, and horn, strong in labour and yet good milkers as well."[131] These "northern"

---

129. ERO, D/AMW 1/273 (1625), made probate 84 days after writing.

130. Markham, *Cheape and Good Hvsbandry*, p. 86.

131. Joan Thirsk, "Horn and Thorn in Staffordshire: The Economy of a Pastoral County," in *The Rural Economy of England: Collected Essays,* edited by Joan Thirsk (London: Hambledon Press, 1984), p. 170.

cattle were regularly found in the south of England.[132] Thus, when Henry Abell of South Hanningfield left "to my deare & loving wife Susan my browne northren Cowe," the color of the animal was worth mentioning perhaps because it was unusual; as Markham noted, northern cows were "generally" black.[133] Other wills focus on individual bovine markings: John Coole of Little Clacton left Margrett Bond "My wives litle daughter a white backt bullocke";[134] Richard Taylor left his daughter Mary a cow "called white belly";[135] Henry Choppin left his brother Richard "one black Kow wth a white face, with one black bullock wth a white face";[136] and Nicholas Clarke of Aldham left Hellen, his daughter, "my white face Cowe that came from Clacton" (that is, from twenty miles away).[137] Three wills mentioned a cow with a white star on its forehead.[138]

The sex of the animals was another means by which bovids were distinguished. Cows, of course, were female; bulls male, and while the former

---

132. Other black bovines are in ERO, D/ACW 9/48 (1623), made probate 24 days after writing; ERO D/ABW 43/191 (1620), made probate 30 days after writing; ERO, D/ABW 45/184 (1623), made probate 14 days after writing; ERO, D/ABW 47/31 (1626), made probate 36 days after writing; and ERO, D/AMW 3/250 (1630), made probate 695 days after writing.

133. ERO, D/ABW 47/208 (1626), made probate 36 days after writing. Another "brownish" cow was bequeathed in ERO, D/ACW 11/162 (1631), made probate 29 days after writing. Cows other than brended and black ones are also found in Essex. According to Markham, red cows originated in Somerset and Gloucestershire, but the wills reveal them as also common in the southeast. See, for example, ERO, D/ABW 44/137 (1621), made probate 44 days after writing; ERO, D/ACW 8/296 (1620), made probate 75 days after writing; ERO, D/ACW 10/107 (1625), made probate 46 days after writing; ERO, D/AEW 17/193 (1624), made probate 642 days after writing; ERO, D/AEW 19/68 (1631), made probate 48 days after writing; and ERO, D/AEW 19/100 (1631), made probate within 191 days of speaking. Markham claimed that "Pyed" cows,—that is black and white, or "speckled" animals ("pyed" refers to both)—originated in Lincolnshire, and these also appear in the large ERO dataset. See, for example, ERO, D/ABW 48/112 (1627), made probate 33 days after writing; ERO, D/ACW 9/239 (1625), made probate 50 days after writing; and ERO, D/ABW 50/227 (1630), made probate 61 days after writing. The latter will is dated 4 March 1603, but "Francis Stanes of Little Baddowe" is recorded as being buried in the churchyard of St Michael, Woodham Walter (as he requested in his will) on 27 March 1630. The date of the will is likely to be a scribal error: 1603 written instead of 1630. Parish Register of St Michael, Woodham Walter, ERO, D/P 101/1/1 (1559–1789); Markham, *Cheape and Good Hvsbandry*, pp. 86–87.

134. ERO, D/ABW 43/74 (1621), made probate 104 days after writing. Another "Bullocke with the whit Backe" is in the will of Gilbert Church of Fyfield, while Andrew Wait of Asheldham left "a whit flanckd Bullock." ERO, D/AEW 17/160 (1623), made probate 45 days after writing; ERO, D/AEW 18/297 (1629), made probate 55 days after writing.

135. ERO, D/AEW 19/76.

136. ERO, D/ABW 49/229 (1629), made probate 98 days after writing. Another white-faced cow is mentioned in ERO, D/AEW 19/238 (1633), made probate 55 days after writing.

137. ERO, D/ABW 52/174 (1634), made probate 45 days after writing.

138. ERO, D/AEW 17/323; ERO, D/AEW 18/32 (1626), made probate 87 days after writing; ERO, D/ABW 49/246 (1628), made probate 11 days after writing.

were regularly given in wills, the latter were rarely bequeathed.[139] Only six bulls appear in the large ERO database: Edward Pond of Margaretting, a husbandman, left his youngest daughter Elizabeth "my Bull wch I have at goodman Amats";[140] John Poole of Downham, a yeoman, left his wife Sarie "two bules";[141] John Greene, yeoman of Terling left his son Roger "one greate bull";[142] and Thomas Porter of Tendring directed his brother-in-law, John Pilgrim, to sell (among other things) "two bulls" in his will.[143] A bull would have been expensive to keep and would have needed a lot of fodder over the winter, and while he was a necessary animal, the need for him was not constant (unlike a dairy cow whose continuous presence was required for the production of milk). Indeed, Mascall proposed that "among a hundred calues two shal be sufficient for to make buls"; and that other male calves should be gelded.[144] Many testators who owned cows would have used the labors of a common bull,[145] or would have used a neighbor's bull to "cover" their cows, perhaps offering a payment toward the upkeep of the animal in return. John Taylor tells a tale of "A Parson of a Countrey Village (for the encrease of the Towne Calves) [who] kept a lusty *Bull*, which serv'd for the use of the whole Parish";[146] and Ralph Josselin recorded in 1651:

> The litle blacke northerne cowe tooke bull. May. 31. Smallbigs bulled. June.2. both by brindle bulls; June 10. morning wessney buld by Mr Littells bull. June. 12. Redbacke bulled: June. 15. Brownbacke: July. 2. morning wesney.[147]

Thus, every dairying household need not include a bull, so it is logical that few would be included in wills, and what seems clear from the lack of bulls

---

139. Leigh Shaw-Taylor's data from Northamptonshire inventories from the first half of the eighteenth century also reveals a small number of bulls. Only 3 percent of the individuals in his sample owned a genitally intact adult male bovid. Shaw-Taylor, "Chapter Three: The Nature and Scale of the Cottage Economy," p. 13, Cambridge Group for the Study of Population and Social Structure, http://www.geog.cam.ac.uk/research/projects/occupations/abstracts/paper15.pdf, accessed 21 July 2016.

140. ERO, D/AEW 16/301 (1620), made probate 116 days after writing.

141. ERO, D/ABW 52/185 (1634), made probate 51 days after writing.

142. ERO, D/ACW 10/66.

143. ERO D/ABW 47/195.

144. Mascall, *Government of Cattle*, p. 49.

145. See Shaw-Taylor, "Chapter Three," p. 13; and Baker, "Stanton-by-Bridge," p. 81.

146. John Taylor, *Bull, beare, and horse, cut, curtaile, and longtaile* (London: M. Parsons, 1638), B4v.

147. Josselin, *Diary of Ralph Josselin*, p. 247. On 13 March the following year, Josselin recorded that "Norden calved a bull calfe shee tooke bull May. 31. 1651. past, which is 40 weekes and 5 dayes." On 3 April, Josselin wrote, "Wesseny calved. shee tooke but twice; last time was. July. 2. she went not 40. weekes full(.) a blacke bull calfe." Josselin, *Diary of Ralph Josselin*, pp. 274 and 276.

in the large ERO dataset (and of uncastrated adult male animals in general[148]) is that the majority of male animals on the farms and small holdings in Essex were sold on or slaughtered before they reached maturity. This is possibly the fate planned for the "2 two yeare old steares" in Fuller's will: he had a "teame of horses" to do the plowing, so the steers could have been intended for the meat trade (although they might have served as a second plow team). Indeed, in contrast to the six bulls in the large ERO dataset, there at least ninety-nine bullocks ("at least" because some wills include general terms—e.g., "a bull-ock," "one of the bullocks," "the best bullock," "bullocks"[149]). This is a figure that seems to point to the business of fattening up male bovids for slaughter.

However, it is not certain that all of the bullocks in the wills in the large ERO dataset are male: as with pigs/piglets, the terminology is not as clear as it is in modern usage. For example, the husbandman Thomas Pennifeather bequeathed "twoo Steere bullocks" to Humfry Smith and two "cow bullocks" (one each) to John Lawrence and John Smith,[150] the Hatfield Broad Oak yeoman Geoffrey Sagger bequeathed his daughter Luce "one heiffer bullocke,"[151]

---

148. As with bulls, rams are rare: only four are bequeathed in the large ERO dataset. These are in ERO, D/AEW19/299 (1634), made probate 50 days after writing (one ram); and ERO, D/ACW 11/142 (1632), made probate 62 days after writing (three rams). Ram lambs are bequeathed in ERO, D/ACW 10/227 (1628), made probate 78 days after writing; ERO, D/AEW 19/141 (1632), made probate 101 days after writing; and ERO, D/APbW 1/15 (1632), made probate 136 days after writing. Wethers (castrated male sheep) are bequeathed in ERO, D/ABW 44/96 (1622), made probate 93 days after writing; ERO, D/ACW 11/142 (1632), made probate 62 days after speaking; ERO, D/ACW 11/198 (1631), made probate 1,262 days after writing; ERO, D/AMW 2/50 (1623), made probate 88 days after writing; and ERO, D/AMW 3/28 (1625), made probate 61 days after writing. Fuller also did not include an adult male pig in his will, and it is difficult from the terminology used to discern who did own such an animal. No boars were bequeathed in the large ERO dataset, but the term "hogs" might include adult male pigs.

149. See ERO, D/ABW 43/28 (1621), made probate within 1,439 days of writing; ERO, D/ABW 43/74 (1621), made probate 104 days after writing; ERO, D/ABW 46/190 (1624), made probate 23 days after writing; ERO, D/ABW 48/132 (1627), made probate 45 days after writing; ERO, D/ABW 48/155 (1627), made probate 51 days after writing; ERO, D/ABW 49/205 (1629), made probate 92 days after writing; ERO, D/ACW 11/95 (1630), made probate 357 days after writing; and ERO, D/AEW 19/182 (1633), made probate 4 days after writing.

150. ERO, D/ACW 11/154 (1632), made probate 73 days after writing. Cow bullocks also appear in ERO, D/ABW 49/9 (1629), made probate 84 days after writing; ERO, D/ACW 10/267 (1627), made probate 98 days after writing; and ERO, D/ACW 11/22 (1628), made probate 236 days after writing.

151. ERO, D/ABW 47/186 (1625), made probate 2,184 days after writing. The *OED* defines "heifer" as "a young cow, *spec.* one that is over one year of age but has not yet calved (or, in some areas, that has not calved more than once)."The *OED*'s parenthetical meaning of heifer can perhaps be traced in the will of Edward Ivet of Thundersley, who left his daughter Sara "one heifer that hathe a calfe." ERO, D/ABW 51/312 (1632), made probate 75 days after writing.

and Thomas Meade, a yeoman of Great Stambridge, left his son "one black Bullocke that is not wth Calfe," implying that he might have possessed a bullock that was with calf.[152] Alice Maskall, a widow, might be reiterating this terminological confusion when she bequeathed her son John "the milch Bullock wch is tow yeares of Age," but it is possible that she was referring to an immature animal that was still sucking its mother's milk, a costly practice (the milk would not be used for human consumption) that would have affected the taste of the animal's meat and thus increased its value.[153]

If we take "bullock" to mean young animal (a usage the *OED*'s definition allows[154]) rather than only young male animal, the apparent confusion about the sex of bullocks in early seventeenth-century Essex wills is explicable, even if the sex of the animals remains unclear. But determining the age of young cattle is also difficult when the terms "calves," "weanels" and "bullocks" are used. In current usage, an animal develops from a calf (still drinking its mother's milk) to a weanel (weaned from the mother) to a heifer or bullock, and finally into a cow, a bull, or a steer. But just as "pig" may refer to old or young swine, the terminology to describe young bovines was not generally agreed upon in this period. The Great Bardfield tailor Thomas Dawson's bequest of "my redde Wennell or Bullocke" to his granddaughter Sara Chapman reveals the difficulty—or perhaps it is evidence that Dawson was hedging his bets about what stage of development the animal would be at when death finally took him.[155] John Sturton, a tanner from Great Dunmow, made the distinction clearer when he bequeathed to his son-in-law William Rayment "my black bullocke that I weaned being now ffour yeares ould."[156] The lack of clarity in terminology may actually, however, reveal different farming methods. When Edward Collen, a husbandman from Thundersley's, bequest of "one of my yearelyng [one year old] bullocks"[157] is read alongside the Upminster yeoman George Sawkins's inclusion of "a calf of a yeere ould" to each of his two sons and "a calf of tow yeer ould" to each of his three daughters,[158] what we might

152. ERO, D/AEW 18/32.

153. ERO, D/ABW 49/349 (1627), made probate 69 days after writing.

154. The *OED* recognizes that "bullock" means "a young bull, or bull calf," but it also notes that the term could be "applied loosely to a bull, or bovine beast generally."

155. ERO, D/ABW 48/197 (1626), made probate 77 days after writing. I think it is unlikely that the bequest offered the legatee a choice of a weanling or a bullock.

156. ERO, D/AMW 3/250 (1630), made probate 695 days after writing.

157. ERO, D/ABW 46/190.

158. ERO, D/AEW 19/186 (1633), made probate 60 days after writing.

be seeing is not necessarily the interchangeable use of "bullock" and "calf" for animals of the same age. We might also read the different terminology as evidence that the two men were engaged in different kinds of production.

According to Thomas Tusser, cows were likely to give birth between "Christmas & Lent," and calves to be weaned in May,[159] and so for Tusser, the oldest age for removing a calf from its dam would be five months. At that point, the calf would become a "weanel." Markham offered a different sense of weaning, however. He wrote that "there are two wayes of breeding" calves, "the one, to let them runne with their Dams all the yeere, which is best, and maketh the goodliest beast: the other, to take them from their Dams, after their first sucking."[160] The problem with the first of Markham's methods for smallholders is that it would make the cow less productive for the dairy. If Edward Collen, who described himself as a husbandman, relied on his cow for milk for his family, then weaning a calf as soon as possible would make sense. In his will he refers to two cows ("one Cow . . . my other cow"), and if this reflected all he had, then allowing one of those animals to suckle her calf for an extended period of time might not be in his financial interest, nor would it serve the nutritional well-being of his family. In such a context, it is possible that he referred to the one-year-old animal as a bullock because it had been weaned early. The other side of animal agriculture might be visible in George Sawkins's possession of (at least) two "yeere ould" calves as well as three "tow yeer ould" ones. Sawkins described himself as a yeoman and, following Howell's reading, his bequest of £10 to each of his five children might offer further insight into what that description meant.[161] Perhaps he was a man with a large enough herd of dairy cows to allow some calves to stay with their mothers for an extended time. As G. E. Fussell noted, Essex was famed for its veal calves in the seventeenth century.[162] Sawkins lived in Upminster and was closer to the City of London than Collen so he might have been trading in high-quality meat animals at the same time as he kept enough animals to supply his own household with milk.

In addition to focusing on the appearance and age of the cattle, a few wills link animals to particular people. Such links sometimes included details about where animals came from and in those details buying and selling becomes visible: "my twoe black milch Cowes that I bought of Mr Balle,"

---

159. Tusser, *Five Hundred Points of Good Husbandry* (1638), pp. 63 and 84.
160. Markham, *Cheape and Good Hvsbandry*, p. 88.
161. ERO, D/AEW 19/186.
162. G. E. Fussell, *The English Dairy Farmer 1500–1900* (London: Frank Cass & Co., 1966), p. 63.

for example.[163] A husband's testamentary description often included the fact that his wife had brought animals into the marriage, and occasionally this had accidentally comic implications, as when Robert Sabin left his unnamed spouse "the two beasts that she brought & ye bullock yt was hir calfe."[164] In addition, wills can reveal previous gifts or legacies: so Rachel Chapman, a spinster from Stock, noted that "whereas there was gyven vnto me by the last wyll and testament of my late father deceased the some of Tenn powndes of good and lawefull money of Englande and one black howed Cowe prysed at three powndes," these and other things were "still remayninge in the handes of Anthony Chapman my brother." She bequeathed them (perhaps pointedly) to her brother-in-law John Silvester.[165] Finally, John Jeppes, a yeoman from Debden, bequeathed to "Elizabeth Jeppes my daughter two bullocks the one black howed and the other a browne motly ~~which when they were calves hir godmother the wyddow Webb did give her~~."[166] The fact that the detail is crossed out (like the crossing through of John Davison's apology to his wife discussed in the introduction) tells us that the scribe considered the detail unnecessary. The physical descriptions of the bullocks were enough for him. The initial inclusion of the history of the animals, however, tells us something about the father's perception of them.

For Jeppes, as for other testators, what we might call the animals' human lineage was part of who the animals were. Like people—my son, my daughter's daughter, the youngest child of my kinsman, and so forth—animals had human lineages. Knowing the history of the animal had a practical utility: as Louise Hill Curth has noted, it was helpful when it came to giving that animal

163. ERO, D/ABW 52/174 (1634), made probate 45 days after writing. See also ERO, D/ACW 8/294 (1620), made probate 133 days after writing; ERO, D/AEW 17/126 (1623), made probate 75 days after writing; and ERO, D/AEW 18/246 (1634), made probate 230 days after writing.

164. ERO, D/AEW 18/109 (1627), made probate 161 days after writing. For other wills that mention wives bringing livestock into the marriage, see ERO, D/ABW 43/168 (1621), made probate 95 days after writing; and ERO, D/AEW 19/173 (1632), made probate 153 days after writing. In another comic outcome of lack of punctuation, yeoman Robert Parker's will states: "to the said John Parker I bequeath More ouer in case as aforesaid my Wife shall marry one of my Cowes." ERO, D/ACW 12/35 (1634), made probate 330 days after writing.

165. ERO, D/AEW 17/80 (1622), made probate 1,968 days after writing. Her father Matthew Chapman's will, which was written and made probate in 1609, specified the bequests she referred to, although his will mentioned "a black howed bullocke of thre years ould." This shows that this animal was eleven years old when Rachel wrote her will in 1617 (it was made probate in 1620). It also signals that this "bullocke" had become a "Cowe." Matthew Chapman also specified that "Anthony shall haue the disposing of Racheills Xl [£10] vnto the age of xxith or day of mariag," so this may be why she had not received the legacy when she wrote her will in 1617. ERO, D/ABW 10/259 (1609), made probate 17 days after writing.

166. ERO D/ACW 9/91 (1622), made probate 76 days after writing.

medical treatment as the husbandman would know "their tendency to suffer from certain disorders and how they responded to treatments in the past."[167] But there might be more to this comprehension of an animal's past than utility. As Wrightson and Levine noted of the goods bequeathed in wills, the careful describing of them was "not trivial": it had meaning. They cite non-self-moving movables in their discussion, stating that "these various goods were their owners' pride . . . [they] were the outward and very visible signs of success among those best placed to take advantage of the opportunities of the age."[168] But what of the self-moving movables? Can the idea of pride still be applied? And if so, is that pride only about the successful gathering of material wealth? For those less able to "take advantage of the opportunities of the age," it might be the ability to pass on what represented the labor of a lifetime that would allow your family to subsist that was important, and this is visible in the heart-breaking nuncupative will of Thomas Boncham of Matching. In it was recorded his wish to leave legacies of material and emotional value:

> I would that my wife should haue all that I haue duringe her lyfe, and all wilbe little ynough: and I give to my eldest sonne a cow. and to my next sonne as good. and to my two yongest sonns all my sheep vppon my grownd, and to my three eldest daughters 20s a peece, and to my youngest daughter a little H heghfer that I bought, theise things he willed his wife to give them that they might hearby remember him their carefull father.[169]

In addition to being evidenced in the accumulation of inanimate stuff, then, pride might also be felt in the possession of animals. Being a "carefull father" need not relate only to Boncham's concern for his children; it might also be associated with his attentive husbandry of his animals. Indeed, the two are inseparable in his will, as it is the animals that are evidence of his "carefull" parenting.

Christopher Fuller's will—to return to where this chapter began—reveals to us not simply details of his human family, his possessions, and his agricultural

---

167. Curth, *The Care of Brute Beasts*, p. 65.

168. Wrightson and Levine, *Poverty and Piety in an English Village*, p. 39.

169. ERO, D/ABW 44/22 (1622). No date is given for the speaking of the will, but Boncham was buried on 27 February 1622 and the will was made probate 8 days later. Parish Register of St Mary the Virgin, Matching, ERO, D/P 411/1/1 (1558–1746).

labors, it also begins to lay bare just how central quick cattle were to the lives of so many in this period. The animals, like the kettles, candlesticks, and beds, were simply possessions. But they were also delineated in ways that allow us to glimpse how they were used (horses for carts, not riding), how productive they were (the shots, pigs, weanels and heifers, and the cheese), what they looked like, and even, for one animal, her name. In addition, we also know that Fuller's life would have centered on caring for and working with his animals, that the well-being of his human household and his animals were inextricably intertwined. In this world, quick cattle were the supporters of the family; were evidence of human care for other humans; were sentient beings with whom people formed working and, I suggest, affective relationships; and could be individuals. And it is to what seems to be the fullest evidence of animal individuality that the book now turns: names.

 CHAPTER 3

Named Partners and Other Rugs

Animals as Co-Workers in Early Modern England

Naming an animal in a will would seem to be the clearest representation of a close relationship with it. What is named is not only unique (a wether with two teeth might be that) it is also individualized in a way that humans are. No one, after all, bequeaths their goods to "my red-haired niece," "my long-legged cousin," or "my one-eyed son: names are (usually) used to specify legatees.[1] Thus, when they are used for animals, names would seem to distinguish livestock in a particularly human and humanizing way and so give (perhaps bequeath) them a special status. In the context of early modern agriculture, it seems plausible that naming an animal with whom a person had spent so much time and for whom they had expended so much physical and emotional energy, would be a meaningful reflection of that person's relationship with him/her (a named animal ceased to be an "it"), and this chapter looks at the use of such animal names in wills. It takes two foci: the naming practices (what names are chosen, what they do, what they don't do, who does and doesn't get them) and the implications of naming for our understanding of life with livestock in early modern Essex.

---

1. There are, of course, occasions when legatees are not given names. Robert Sabin did not include his wife's name (ERO, D/AEW 18/109 [1627], made probate 161 days after writing), and Thomas Stevens did not name his sister (ERO D/ACW 10/214 [1626], made probate 137 days after writing).

The limitations of using wills as evidence for the practices of naming are substantial, however, and need to be addressed. In the 451 wills in the large ERO dataset that include specific animal bequests, there are only five named horses and twenty-seven named cows (no other species is given a name). These animals appear in fifteen wills.[2] That is, only 0.3 percent of all wills have named animals in them. Taking phrases such as "one of my Cowes," or "my best Cowe save one" to conservatively estimate the existence of just two animals, the total number of cows owned by testators that can be confirmed in the large ERO dataset is 721.[3] That means that less than 4 percent of dairy cows in wills are given names. The true proportion is likely much lower than that, however, as not only am I underestimating what phrases such as "one of my Cowes" might mean I am also excluding from my calculation any cows who might be included in the collecting phrases at the end of wills.

These figures are worth pausing over as they seem to contradict what historians who have looked at the issue have assumed to be the reality of life with cows in the period (and dairy cows are the focus of this chapter). Keith Thomas, for example, argued that cows were "always" given names in early modern England, and Virginia DeJohn Anderson also claimed that some of the animals on even enclosed farms had names (enclosed farms were much bigger than the smallholdings that seem to be visible in many of the Essex wills). For Thomas and Anderson, naming signals the close intertwining of humans and their animals: Thomas wrote that names are "suggestive of an affectionate attitude on the owner's part," and Anderson noted that the physical and emotional closeness of humans and animals from which naming emerges could "undermine the supposedly rigid boundaries between humans and animals."[4]

---

2. Two wills include six named cows each (ERO, D/AEW 19/138 [1632], made probate 28 days after writing; and ERO, D/APbW 1/15 [1632], made probate 136 after writing) and one includes four (ERO, D/AEW 19/76 [1631], made probate 144 days after probate). The length of time between the date of writing and probate of the latter two wills raises the issue of how far in advance testators felt able to write wills that included specific animal bequests. I could find no record of the testator of D/APbW 1/15 (John Grant, a yeoman of Southchurch) to confirm the date when he died. Richard Taylor, a husbandman from Stanford Rivers (D/AEW 19/76), however, was buried on 27 January 1631, thirty-five days before the probate date. Parish Register of St Margaret, Stanford Rivers, ERO, D/P 140/1/1A (1558–1745).

3. The phrases come from ERO, D/ACW 12/35 (1634), made probate 330 days after writing; and ERO, D/AMW 1/267 (1634), made probate 889 days after writing.

4. Keith Thomas, *Man and the Natural World: Changing Attitudes in England 1500–1800* (London: Penguin, 1983), p. 96; Virginia DeJohn Anderson, *Creatures of Empire: How Domestic Animals Transformed Early America* (Oxford: Oxford University Press, 2004), p. 90. Anderson describes the livestock on enclosed farms as "a couple of dozen cows, half as many pigs, maybe a horse or two."

Like Thomas and Anderson, I think dairy cows, at least, were given names in this period, so I would suggest that the lack of animals with names in the large ERO dataset should not be understood as a reflection of reality, and other kinds of evidence would seem to support this. In his spiritual diary, for example, Ralph Josselin gave his cows names in his record of their bulling in 1651 (they were Smallbigs, Wessney, Redbacke, Brownbacke and Norden),[5] and in his early seventeenth-century song "Jack and Joan they Think No Ill," Thomas Campion represented the happiness of the ageing couple in terms of their faith, their participation in country feasts, their love for their son and daughter, and their relationship with their animals, which is represented in the line "Joan can call by name her cows."[6] Here, trans-species intimacy is a marker of rural stability. In this context, it seems that the rare inclusion of animals with names in wills is a reflection not of people's understanding of the nature and status of their animals but of the legal document. A will for many, perhaps, was perceived to be from a world distinct from that of the field,[7] and it is possible that testators censored themselves even before they dictated their testamentary wishes to a scribe.[8]

To name a cow in a will might have been to reveal oneself to not recognize the difference between separate social realms, with the evidencing of a close relationship with animals in a will reflecting a discursive failure on the part of the testator. In his *Meditations . . . written for the instruction and bettering of youth, but, especially of the better and more Noble* of 1612, for example, Anthony Stafford advised that "The basest griefe of all, is that, which receiues his birth frō the death of a Horse or a Cow; or from the losse of [gold and silver] the two too high-priz'd Metalls."[9] Here animals were viewed as being of financial, not affective, value and grieving for them figured as an absurdity—the

---

5. The "litle blacke northerne cowe" is the only one not named in the original list, but in the margin Josselin wrote the name "Norden." I return to this point later. Ralph Josselin, *The Diary of Ralph Josselin, 1616–1683*, edited by Alan Macfarlane (Oxford University Press, 1976), p. 247.

6. Thomas Campion, *Two Bookes of Ayres* (London: Tho. Snodham, [1613?]), G2r.

7. This is a focus of my essay "Farmyard Choreographies in Early Modern England," in *Renaissance Posthumanism*, edited by Joseph Campana and Scott Maisano (New York: Fordham University Press, 2016), pp. 145–66.

8. Illiterate people may have had particular reason to self-censor: someone else (often the local vicar) had to write their will for them, and the legally required witnesses to it who were also illiterate had to have the will read aloud to them. As Ralph Houlbrooke has noted, "The move away from oral confirmation of a publicly read document to reliance on ratification by witnessed signature and seal facilitated the move to privacy." Before then, the last will and testament might become public knowledge before the death of the testator, a factor that might affect how they bequeathed their goods. Houlbrooke, *Death, Religion, and the Family in England, 1480–1750* (Oxford: Oxford University Press, 1998), p. 109.

9. Anthony Stafford, *Meditations, and resolutions, moral, divine, politicall* (London: H. L., 1612), p. 98.

application of an emotional response to what was a business issue. Such a view did not consider animals as sentient beings but as mere instruments for use, the equivalent of other kinds of valuable things, such as precious metals. Stafford's assertion, by implication, is rejected by Thomas Dekker's North-Country-man in a dialogue about the impact of the "great Snow" from 1614. Speaking to the Citizen, the North-Country-man notes:

> All your care is to prouide for your Wiues, Children, and Seruants, in this time of sadnesse: but Wee goe beyonde you in cares, not onely our Wiues, our Children, and household Seruants, are vnto vs a cause of sorrow, but wee grieue as much to beholde the miserie of our poore Cattell (in this frozen-hearted season) as it doth to looke vpon our owne Affliction. Our Beastes are our faithfull Seruants, and doe their labours truly when wee set them to it: they are our Nurses that giue vs Milke; they are our Guides in our Jornies; they are our Partners, and helpe to inrich our State: yea, they are the very Upholders of a poore Farmers Lands and Liuings.[10]

In the face of this declaration of shared sentience and partnership, Stafford's perspective can be said to have emerged, you might say, from the realm of books rather than work, to have been written in the study rather than the yard. And the law, sometimes viewed as a malign external force, seemed to replicate such a perspective when it made clear that the personal names of animals were of no value.

Thus, where Henry Swinburne argued that misnaming a human legatee in a will could mean that "the disposition is voide," no such fate occurred to the bequest in which an animal was misnamed:

> The error of the testator in the *proper name* of the thing bequeathed, doth not hurt the validitie of the legacie, so that the bodie or substance of the thing bequeathed bee certaine: for example; the testator dooth bequeath his horse *Bucephal*, whereas the name of his horse is *Arundell*: this errour is not hurtfull, but that the legatarie may obtaine the horse *Arundell*, if the testators meaning be certaine: for names were deuised to discerne things: If therefore wee haue the thinge, it skilleth not for the name.[11]

---

10. Thomas Dekker, *The cold year. 1614. A deepe snow: in which men and cattell haue perished, to the generall losse of farmers, grasiers, husbandmen, and all sorts of people in the countrie; and no lesse hurtfull to citizens* (London: W. W., 1615), B2v.

11. Henry Swinburne, *A Brief Treatise of Testaments and Last Willes* (London: Iohn Windet, 1590), 244r–245r.

On the other hand, misnaming the species ("the *name appellative*" as Swinburne terms it), like misnaming the legatee, leaves the legacy "voide." He wrote:

> The reason of the difference (I meane of the diuers effects betwixt the error in proper names, and the error in names appellatiue) is because a proper name is an accident, attributed to some singular or indiuiduall thing, to distinguish the same frō other singular things of the same kinde: whereas names appellatiue doo respect the substance of things, and being common to euerie singular of the same kinde, make them to differ from things of other kinde or substance.

So "the testator intending to bequeath a horse [who] doth bequeath an oxe" will find their bequest void.[12] Giving an animal its proper name was thus an indifferent matter—it carried no weight. It was the species of animal that had the legal standing, not the individualized creature. For the law, it seems, one cow was as good as another, so when it came to writing a will, perhaps clarity was felt to be best exercised through description rather than naming if a specific animal was intended.

In addition, just as the law did not recognize animals' names, so it also presented quick cattle in a particular way that transformed knowing them into a question of superficial legibility that could be enacted with a head count rather than through an intense one-to-one engagement. It transmuted them, indeed, from Smallbigs, Wessney, Redbacke, Brownbacke and Norden, as Josselin records his cows in his record of his daily life with them, into "5 cowes at 5li. round [i.e., each]," as they are represented in his inventory of his "outward estate" in March 1651.[13] According to the law, cows, like other stuff, were ownable: or, as the Elizabethan lawyer William Lambarde put it in 1592, they were stealable, and their stealability was what marked them as property. Lambarde wrote:

> Money, plate, apparell, housholde-stuffe, Corne of any sort (or haie, or fruit) that is seuered from the ground, horses, mares, coltes, oxen, kine, sheepe, lambes, swine, pigges, hens, geese, ducks, peacockes, turkies, and other beasts, and birds of domesticall (or tame) nature, are such, as felonie may bee committed in the taking of them.[14]

---

12. Swinburne, *A Brief Treatise of Testaments and Last Willes*, 245v–246r.

13. Josselin, *The Diary of Ralph Josselin*, pp. 247, 240.

14. William Lambarde, *Eirenarcha: or the office of the iustices of peace in foure bookes* (London: Ralph Newbery, 1592), p. 267.

The crucial phrase for animals is "of domesticall (or tame) nature." In his 1618 advice manual for country justices, Michael Dalton followed Lambarde and noted that domestic animals were like all other goods and so were "absolute property," and thus their legal status was unproblematic. Stealing a sheep was the same as stealing "Money, plate, apparell, [or] household-stuffe." With wild animals it was more complicated, however. They could be deemed legal possessions "by making them tame" and only for "so long as they remaine tame."[15] William Noy, who was attorney-general from 1631 to 1634, explained in more detail what "remaining tame" might mean:

> propertie is not in any [wild animal], not after they are made tame, longer then they are in his Possession; as my Hounds following me, or my man, or my Hawke flying after a foule, or my Deer haunting out of my Park. But if they stray of their own accord, it is lawfull for any man to take, and the heire shall have them.[16]

What Noy recognized was that while "wild" animals such as hawks and deer could be trained or tethered or enclosed within specific spaces, they also possessed the capacity to "stray of their own accord"; that is, they had volition. However friendly, however obedient, however penned in, some animals were perceived to be always potentially wild, and that potential was enough to classify them as ownable only in "Qualified" or "Possessory" ways. Such ownership could be "attained," but it could also be lost.[17]

Noy's outline also makes clear why there are no dogs in the large ERO dataset. "Hounds" were classed as wild animals, and their tameness was the only thing that created their status as property. When a hound followed the owner, it was owned, but, by implication, that qualified status became

---

15. Michael Dalton, *The Covntrey Ivstice, Conteyning the practise of Ivstices of the Peace out of their Sessions* (London: Society of Stationers, 1618), p. 234. I have discussed the legal conception of wild and domestic animals in more detail in *Perceiving Animals: Humans and Beasts in Early Modern English Culture* (Basingstoke: Macmillan, 2000), pp. 125–37.

16. William Noy, *A Treatise of the Principal Grounds and Maximes of the Lawes of this Nation* (London: T. N. for W. Lee, 1651), pp. 51–52. "Haunt," according to the *OED*, can mean "to use or employ habitually or frequently; *refl.* to use, accustom, or exercise oneself." The customary nature of the action is legally significant, I think, and the reflexive nature of the verb is also important because it signals self-awareness. The *OED's* first recorded usage of the term for a mental action dates from 1615, when Thomas Jackson referred to "reflexiue, or examinatiue acts of the vnderstanding" in *Iustifying faith, or The faith by which the just to liue* (London: Iohn Beale, 1615), p. 8.

17. Dalton notes that "when a man have beasts or fowles (that be sauage and in their wildenesse) . . . by reason of a Parke or Warren, &c. (as Deere, Hares, Conies, Phesants, or Partridges, or the like which be things of Warren) he hath no propertie in them . . . but they belong vnto him *ratione Priuilegij* (for his game and pleasure) so long as they remaine in the place priuileged." It is for this reason that stealing such wild creatures from parks or other restricted sites is poaching and is thus a crime. Dalton, *The Covntrey Ivstice*, pp. 234–35.

impossible once the owner had died. Not only did the owner cease to exist as an owner, of course; but by extension, at the death of the owner, the dog, freed from the bounds of ownership, had to be re-tamed (attached by evidence of loyalty to another person). This process was not recognized as automatic, so ownership of dogs was based on a conception of taming that was qualified and transient. "Beasts, and birds of domesticall (or tame) nature," on the other hand, were not expected or required to themselves evidence their being owned: they did not need to do anything to show whose property they were. These creatures could be transferred to a new owner with only a bill of sale or a legal will.[18] Thus, just as the individual name is indifferent and the species crucial in a will, so the law presented quick cattle as lacking possession of what might be deemed a character. They were things that could be transferred, whereas a hound or a deer could turn away and needed to be constantly reclaimed, trained, tethered.

Some pets, clearly, could be characterized as being thoroughly tamed, of course, but this did not mean that the early modern law regarded them as possessions. Lambarde wrote, "to take dogges of any kind, apes, parats, singing birds, or such like (though they be in the house), is no *Felonie*: because these latter bee but for pleasure onely, and are not of any value."[19] Here the crucial category is "pleasure." An animal's raison d'être—as the early modern law represented it—was to have a use, and recreation was the antithesis of that. To train a dog in order to use it to exercise one's soldierly qualities in practice for war (a reason given for hunting) was to invest it with legal value.[20] To tickle it, stroke it, or play with it, on the other hand, was merely recreational, so was deemed legally meaningless: like an animal's proper name, "it skilleth not." Indeed, the possession of "useless" dogs was not just a problem for the law. Nicholas Breton's Mad-cappe, for one, bewailed the decline of life in the early seventeenth century, citing among many of the things that he mourned

---

18. See, for example, the receipt for "ffortie draught oxen one Cow collour black ffortie ounces of gilte plate a longe Cloake of Townye broad Cloth a lether jerkyn thick lace wth silver lace faced wth Cloth of silver and skertes lyned therewith a skye Cullored Cloake of Spanishe selke wth whood arme bases trymmed with silke lace & tuffed with silke a payre of black velvett hose wth payres thick laced & tyed wth taffeta some nett drawn forth of satten Cutt vppon taffita Sarcemett and silver buttons and all that lumpe of haye lyenge in the stable at Hackney." "Receipt from William Binder and Edward Hall to John Danyell of Hackney, Middlesex," NA, WARD 2/55A/191/2 (1600).

19. Lambarde, *Eirenarcha*, p. 268.

20. Thomas Cockaine, for example, wrote that his "first commendation of Hunting" was that "hunters by their continuall trauaile, painfull labour, often watching, and enduring of hunger, or heate, and of cold, are much enabled aboue others to the seruice of their Prince and Countrey in the warres, hauing their bodies for the most part by reason of their continuall exercise in much better health, than other men haue." Cockaine, *A Short Treatise of Hunting* (London: Thomas Orwin, 1591), A3r–v.

from the idealized past was the fact that back then "no man kept a dogge but for an vse":

> The *Mastife* chiefly, for to hunt a hogge,
> The *Hound* to hunt the *Hare* out of her mewse,
> And for a piece, a fetching water-dogge,
> Or for to beate a *Foule* out of a bogge.[21]

Sir John Harington's declaration that his dog Bungey "onely fed my pleasure, not my purse" reveals the distance between these two views perfectly.[22] Pets were useless.

A further problem with pets can be traced in the idea embodied in Bungey's owner's father's translation of Cicero's *De amicitia*: true friendship, John Harington wrote, was "neuer vnreasonable."[23] This sense of the separation of reasonable from unreasonable—and thus human from animal—was also reflected in the law's refusal to allow pets the status of ownable beings. "Though they be in the house," these animals that "onely fed my pleasure, not my purse" were not proper comrades, and because of their uselessness neither were they even the equivalent of flockbeds, candlesticks, stained cloths, or pewter platters, objects that could be valued and bequeathed. Rather, they were (according to a law coming down from the Romans) not things, but nothings. In legal terminology, pets were *"res nullius"*—nobody's property—which meant that the *res* could be taken by the first person who came along with no felony being committed. This, in turn, meant that conceptions of loyalty, friendship, and affection were not legally recognized.[24] Being "nobody's property" was thus

---

21. Nicholas Breton, *Old Mad-cappes new Gally-mawfrey* (London: Richard Iohnes, 1602), D1r.

22. John Harington, "*Against Momus, in praise of his dogge Bungey*," in *The Most Elegant and Witty Epigrams of Sir John Harington, Knight* (London: G. P. for Iohn Budge, 1618), Iv. This, of course, is not to say that the animals in the home did not elicit feelings in those who cohabited with them in the early modern period. As early as 1521, one judge argued that just as "special property" could be had in "deer in my park, or fish in my pond. . . . so it is as to a tame beast which I use in my house. . . . For if I have a singing bird, though it be not pecuniarily profitable, yet it refreshes my spirits and gives me good health, which is a greater treasure than great riches. So if anyone takes it from me he does me much damage for which I shall have an action." Here the animal's value is not just financial. However, the judge's statement is not about the nature of the animal itself; rather, it is (like the anti-cruelty ethic) utterly self-centered. The bird refreshes the judge's spirits, and it is those spirits that should be protected by the law. By implication, the bird, he argues, should be the legal equivalent of a painting that pleases him or a musical instrument the playing of which keeps him entertained, and stealing such an animal should allow for more than a prosecution for trespass. J. Brooke quoted in W. S. Holdsworth, *A History of the English Law* (London: Methuen, 1925), 7:489.

23. Cicero, *The Booke of freendeship of Marcus Tullie Cicero*, translated by John Harington (London: Tho. Berthelette, 1562), p. 16r.

24. The Roman origin is mentioned in Morgan and Another, Executors of John, Earl of Abergavenny, Deceased v William, Earl of Abergavenny, 10 December 1849, 8 Common Bench Reports 768

not a claim to innate dignity or worth, but to its opposite: with no owner, the pet was felt to have no inherent value.

Quick cattle were thus caught between two distinct legal conceptions of animals. Unlike pets, they were ownable (and stealable) objects and so had a status in the law that allowed them to be bequeathed in wills. But these creatures "of domesticall (or tame) nature" were also legally understood to lack the volition possessed by wild animals because they were not perceived to have the capacity to escape: the "absolute" nature of legal ownership of domestic animals precluded this. This was made clear in the law of *levant et couchant* that was defined by the early sixteenth-century barrister and printer John Rastell in the following terms:

> Leuant, & Couchant is sayde, when the beastes or Cattell of a straunger are come into an other mans ground, & there haue remayned a certen good space of time so longe that they haue wel fedde, & also rested them selues.[25]

The name of this law, *levant et couchant*—standing up and laying down—reflects the time necessary for such animals to have "wel fedde, & also rested them selues." Once this had happened, the animals could be distrained (withheld)—the term distressed was also used—until the owner paid a fine and/or settlement for the damages the animals had caused to the invaded person's property.[26]

Because ownership of them was absolute, the responsibility for distrained animals remained the owner's (not the distrainer's) so long as they were held on common ground or

> in an others ground so that [the distrainer] geeue notice to the [owner], that hee (if the distresse be a quicke beast) may geeue to it foode, and then if the beast dye for defaut of foode, hee that was dystrayned shall bee at the losse, and then the other may distrane agayne for the same rent or duitie.[27]

137 E.R. 710. James Manning, T. C. Granger, and John Scott *Common Bench Reports: Cases Argued and Determined in the Court of Common Pleas in the Trinity Vacation, and Michaelmas Term and Vacation, 1849*, vol. 8 (Philadelphia: T. & J. W. Johnson & Co., 1865), 767–98.

25. John Rastell, *An exposition of certaine difficult and obscure words, and termes of the lawes of this realme, newly set foorth and augmented, both in french and English, for the helpe of such younge students as are desirous to attaine knowledge of ye same* (London: Richardi Tottelli, 1579), p. 145r.

26. This term was used in a slightly different way in the deathbed statement of William Nevell, a glover from Sandon, who gave his daughter Joane 20s a year for life to be paid by her brother William Nevell "and his heires out of his land and for non payt therof she is to take a distresse of such cattell as are thervpon going vntill she be pd." ERO, D/ABW 49/238 (1628), made probate within 50 days of speaking.

27. Rastell, *An exposition of certaine difficult and obscure words*, p. 74v.

Thus, while domestic animals could wander into the wrong field if a gate was left open, for example, ownership of them was not challenged. Instead, what happened was that the careless owner who had allowed them to wander was liable for the damage they caused. This makes these creatures categorically different from Noy's deer that strayed of its "own accord," as domestic animals could not revert to the ownership of the possessor of the land they were on, nor could they become wild, because they were always already domestic. For this reason, they were only ever temporarily withheld (distrained) from the owner's possession. In relation to dairy cows, this might mean, although the law does not state it, that the milk produced during the period of distraint would become the property of the distrainer,[28] or that the financial loss and ill health that was a consequence of a failure to milk distrained (or perhaps here, more aptly, "distressed") cows was punishment of the human owner (the law would not concern itself with the impact of the cows' suffering on the cows). In legal terms, domestic animals were perceived to never be able to escape; they could simply "come into an other mans ground." Their status was fixed and possession of them was absolute—not altered or qualified by changing circumstances. For this reason, "distraining" did not refer to a removal of training (dis-training) because if that could happen, it would challenge the absolute ownership of such animals. Rather, "distraining" used "train" to mean "lead astray" and seems to suggest the curtailing of the animals being further led astray.[29] Thus, while the law of *levant et couchant* emphasized the property status of the animal, it also (and consequentially) underlined the legal fact of the animal's own lack of true volition. Quick cattle were self-moving (quick), to be sure, but their lack of will meant that they were legally just stuff (cattle) that (could be) moved around.

This brief foray into the legal status of domestic animals allows us to understand why some animals and not others are found in wills in the early seventeenth century: why, in particular, domestic pets are absent from the large ERO dataset. In addition, and more significantly here, it reveals the gap between the legal definition and the lived relation. Wills were always already pulling in two directions: on the one hand, the legal status of domestic animals allowed a testator to bequeath what were simultaneously valued, valuable, and invaluable things to their kin, but on the other hand, that same legal status made it impossible for the law to fully reflect what people experienced

---

28. In Thomas Lodge and Robert Greene's 1590 play, the poor man who has "pawne[d] my good Cow" states that "the cowes milke [was] for vsurie"; i.e., the cow's value covered the value of the loan and her milk covered the interest. Lodge and Greene, *A Looking Glass for London and England* (London: Barnard Alsop, 1617), Dr.

29. See *Oxford English Dictionary*, s.v. "train."

in day-to-day life with livestock. There is a shift, as Josselin's diary shows, between counting and being with. Testators were constrained (I am tempted to say distrained—stopped from being led astray) because of the limitations placed on the way livestock were perceived and represented, and yet they had to work within those constraints to fulfil their testamentary wishes. Naming an animal in a will marks, if you like, the point at which the split between these two perspectives is made most apparent. It gives us the clearest glimpse of the disjunction between life and the law.

But there are other reasons for taking seriously the issue of naming animals. First, what is clear from even the small number of wills in which testators in Essex did name their livestock is that how animals were named was a complex matter. Second, in the complex practices owners used to name dairy cows can be traced aspects of life with animals that might otherwise go unnoted, and what is revealed points us to think about the fields and yards of early modern England not only as shared spaces but as workplaces that were peopled with both humans and animals. How those humans and animals worked together tells us a great deal about the nature of their relationships, about both physical and emotional engagement. A look at a couple of wills will give a fuller sense of naming practices, and this, in turn, will lead us back to day-to-day lives in early modern Essex.

In his will, which was written in December 1623 and made probate in April 1625,[30] William Walford senior of Cold Norton, a village ten miles southeast of Chelmsford, left his wife Ursula the lease of their house and land in Cold Norton and:

> the three poundes that is in Reddy money in my howse and seaven
> of firkins of ^ye^ butter wch is likewise in the howse ^toward ye^ ~~to~~
> paymt of ye Rent wch is now dew and shalbe dew at o'r lady next [i.e.,
> 25 March] to my landlord Edwarde Bailie Item I give to my said wife
> all my howshould stuffe, and all my kyne except two the one Called
> Evered and the other Called the Bullocke New Come, Item I give to my
> said wife my xxiii Ewe sheepe and one Rame, and also my sowe, Item
> I give to my kinsman William Walford of Cold Norton ye younger my
> lease of the howse & land belong ing to the Towne of Maldon Called
> Sabernes or by what other names soever it is called.

---

30. This may be the same month he was buried. The page of records for 1625 in the parish register of St Stephen, Cold Norton, is very faded, but it is possible that someone with the family name Walford was buried on 2 April 1625. ERO, D/P 379/1/1 (1539–1771).

In addition, he gave 40s to his sister Annie Mott and 10s to his kinsman Thomas Mott, both to be paid at Michaelmas, and 6s 8d to the poor of Purleigh.

> And lastlie I make the said William Walford my kinsman my executor desireing him to paie my debtes to see me decently buried And this my last will pformed. In wittnes whereof I have herevnto put my hand and seale the daie and yeare above written.

William Walford then signed his name.[31]

This will raises a number of issues related to naming practices that begin to reveal the complexity of what might seem to be a straightforward marker of the testator's assessment of an animal's individuality. First, and obviously, the will makes clear that even for humans a name is not always enough to mark individuality. The term "senior" included after the testator's name, like the use of "old" in relation to Christopher Fuller's cow, represents Walford's age in comparative terms, here in relation to the other William Walford of Cold Norton, who is named in the will as both his kinsman and "ye younger." But it also reveals that if the name "William Walford" was used on its own, the legacy might be "devoide" because the practice of using established family names and the possibility that a family would remain in the same place over generations makes it possible that an individual would share the same name with someone else in the same location. Christopher Fuller's eldest son was called Christopher, there are two William Walfords. Indeed, in both Cold Norton and West Bergholt during the period of the large ERO dataset, over one-third of male infants baptized in the local church were called by the same first name as their father.[32] For this reason, an additional adjective might be needed to distinguish humans. The name on its own does not individualize. William Walford's being termed "senior" thus reminds us that a name even

---

31. ERO, D/ABW 46/107 (1625), made probate 494 days after writing. All seven wills from Cold Norton during the period 1620–1634 are included in the large ERO dataset. Walford's is the only one that includes specific bequests of animals.

32. In the fifteen-year period of the large ERO dataset (and noting the illegibility of the page for 1625), 13 male (of 36) and 3 female (of 31) infants were baptized with a parent's name in Cold Norton (the figure for females is lower in part because mother's names were often not recorded in the registering of the baptism). ERO, D/P 379/1/1. In West Bergholt, of the 79 male children baptized in the same period, 29 were given their father's name and of the 75 female infants baptized, 17 were named after their mothers. Parish Register of St Mary the Virgin, West Bergholt, ERO, D/P 59/1/1 (1559–1658). Because of infant mortality rates, not all of these babies survived childhood. Christopher Fuller's son and legatee William had three sons baptized with the name of William. Clodagh Tait notes that "in England in the late sixteenth and early seventeenth centuries, the names William, John, Thomas, Elizabeth, Anne and Mary accounted for nearly half of all names for men and women respectively." Tait, "Namesakes and Nicknames: Naming Practices in Early Modern Ireland, 1540–1700," *Continuity and Change* 21, no. 2 (2006): 315.

for a human is a marker of more (or perhaps that should be less) than just individuality.[33]

A second issue related to naming is revealed in the two animals Walford refers to by name in his will. The "kyne," Evered and New Come, are excluded from a bequest and then are never mentioned again. This seems odd—especially for animals so clearly individualized.[34] However, in law, by default, they would have been left to the executor, William Walford the younger, as the executor came into possession of everything not bequeathed except entailed land and property and distributed it as the will required. He or she kept what was not for distribution.[35] What the invisible bequeathing of Evered and New Come tells us is that named animals might be included in the residue of unbequeathed goods in the collecting phrase at the end of a will.[36] This might seem to place them among the leftovers and not as specially individualized creatures and thus to work counter to how we might expect a name to function. But there is an alternative reading here: Walford's will reminds us that a named cow might not be a special cow, or, to put it another way, that all cows might have names and that when names are used in wills it is simply to specify the ones being bequeathed.[37] In this reading, the exclusion of Evered and New Come does two things: it allows us to read the collecting phrases as not simply reflecting the belief that "cattle" were anonymous quick goods, and it allows us to see, apparently paradoxically, that naming might not be assigning special status to a cow, because, as Keith Thomas argued, all cows had names.

The third issue Walford's will raises centers on the cow named New Come (or, if it is her full name, the Bullocke New Come). I am assuming that she was female for a number of reasons: first, we have seen that "bullock" was used in wills to refer to both male and female animals; second, Walford's will lists

---

33. The issue of the family name might also be at stake in the naming of property. In this will, the property in Maldon has a name, Sabernes, but Walford then adds "or by what other names soever it is called." This phrase or one like it is often seen in wills that include named property and perhaps signals that the name of a property was not fixed and could be changed across time with different owners.

34. Another will shows how animals that were excluded from one bequest might appear in another. Hester Woulfe, the Rickling widow who left the sow with piglets, bequeathed to her daughter Alyce Mills "all my fowles about the yard, excepting thre pullets" and left Jhoane Mylls "my dawghters Dawghter . . . thre pulletts before excepted." ERO, D/ABW 48/52 (1627), made probate 56 days after writing.

35. See Swinburne, *A Brief Treatise of Testaments and Last Willes*, pp. 114r–116v.

36. Walford does not include a collecting phrase in his will, but it was legally unnecessary because property that had not been specifically bequeathed would automatically have gone to the executor.

37. The same might be said of the weaver William Younge's testamentary documentation. The inventory of his goods included three cows, while his will included only the bequest of "my heiffer called Nan." It is likely that Nan was named not because she was a special cow but because she was a special bequest. ERO, D/AMW 2/125 (1628), made probate 45 days after writing.

New Come as one of "all my kyne except two"; and third, because only female bovine animals are named in all other wills. But New Come is not only what this bullock is called. I would suggest that it also marks that she might have been at some time the newest arrival on Walford's property. Here, the name could be more than simply a name; it could be a comparative term. "New Come" would, of course, come to function unproblematically as a proper name as it got used and as the animal continued to be called New Come after she ceased to be the newest arrival in the herd. But in its initial usage "New Come" reminds us that in animal naming practices in the early modern period, names could be less distinct from descriptions than might appear.

Fourth, while Walford includes other animals than Evered and New Come in his will—"my xxiii Ewe sheepe and one Rame, and also my sowe"—none of them is given a name. Following the logic of the second point above, this does not, of course, automatically signal that these animals did not have individual names. Yet no testator in any will in the large ERO dataset named anything but horses or cows.[38] Pigs, we have seen, can be understood as temporary possessions: owned for a short season before being slaughtered for meat. However, Walford's pig is "my sowe," and this description signals, first, that she is singular (she is not "one of my sows," for example), and, second, that she might be his breed animal and so have a longer-term value: Gervase Markham wrote that "A Sow will bring Pigs from one yeere old till she be seauen yeeres old."[39] Whether such a creature might be given a name is not something the wills provide evidence for (but absence of evidence is not, of course, evidence of absence). The twenty-three sheep are bequeathed en masse, which might reflect people's relationships with these animals. As discussed in the previous chapter, few sheep are given individual descriptions in wills, and alongside the absence of any named sheep, this makes it possible to suggest that sheep did not get names in this period (but, once again, absence of evidence should not be taken as evidence of absence). Like the bequest of the sow, the bequest of "one Rame" is singular, but "one" is different from "my" and opens up the

---

38. Of note is the fact that while the named bovine animals are female, the five named horses all could be male. Stephen Kingstone, a blacksmith from Thorrington, bequeathed to his son Stephen "my horse called Bucke'"; ERO, D/ABW 51/171 (1633), made probate 1,289 days after writing. Steven Amis of Ramsey bequeathed to his son Steven "one Gelding called the buckk Gelding"; ERO, D/ACW 10/266 (1627), made probate 252 days after writing. Roger Hull, a miller from Great Maplestead, bequeathed to his son John "my horse called Clubbe" and to his son Roger "my horse called Jack"; ERO, D/AMW 1/273 (1625), made probate 84 days after writing. Christopher Bufford, a yeoman of Ingrave, bequeathed to his wife Elizabeth "my Nagg called Lock"; ERO, D/ABW 43/178 (1620), made probate 1,465 days after writing. A nag could be female.

39. Gervase Markham, *Cheape and Good Hvsbandry: For the well-Ordering of all Beasts, and Fowles, and for the generall Cure of their Diseases* (London: T. S. for Roger Jackson, 1616), p. 124.

possibility that Walford had two or more such animals and that he was leaving just one of them to his wife and giving the others (by default) to his executor William "ye younger" (although this is perhaps unlikely, as only four rams are bequeathed in the whole of the large ERO dataset). "One" might also signal an indifference to specificity, implying that one ram is as good as another, that, as Swinburne noted, it is the species that is significant and not the particular creature. Whether this means that no ram had a name in early modern Essex is, of course, again, impossible to say. But what Walford's bequest of his sow, ewes, and ram does signal is that it is possible that—for him at least—cows were regarded as nameable in a way that other species were not. Neighborly closeness does not really offer a solution here because like the cows, the sow would have been a regular presence in his yard. But the closeness to a dairy cow would have been very particular: someone—likely a woman—would have leaned against the animal's flank day-in, day-out for years, and maybe this very physical closeness (differently manifested but also present in the act of riding a horse, perhaps) allowed for a distinct emotional bond.

Another couple of documents from a few years after Walford's will and sixteen miles southwest of Cold Norton underline the complexity of naming animals in early modern wills. The trail begins with one of only seven wills in the large ERO dataset from Childerditch, a "sparsely populated" village three miles from Brentwood.[40] The will of the yeoman John Thresher was written in August 1629, and made probate 705 days later on 11 July 1631, six days after he was buried.[41] He left

> vnto my welbeloved wife Phoebe Ten Milche Cowes to be chosen by her self out of such kyne as I shall die possessed of; or ffyftie pounds in money att her Choyse, my best Gelldinge or nagge: And ffive quarters of my best wheate either severed or growinge wherof I shall dye possessed: And all such my housholdstuffe as I shall die possessed of: and not herein bequeathed.[42]

In addition, he bequeathed to her seven acres of copyhold land. His nephew, Nicholas Thresher, received a copyhold tenement with eleven acres of land,

---

40. "Parishes: Childerditch," British History Online, http://www.british-history.ac.uk/vch/essex/vol8/pp17-24, accessed 2 February 2016. Two Childerditch residents' wills are excluded from the large ERO dataset: ERO, D/AEW 18/25 (1626) is damaged and ERO, D/AEW 18/240 (written 1628) has no probate date. Except for Thresher's wife's will—discussed below—none of the other five Childerditch wills in the large ERO dataset include animals, although John Totman, a yeoman, bequeathed his godson Roger 6s to buy himself a lamb. ERO, D/AEW 17/287 (1625), made probate 41 days after writing.

41. Parish Register of All Saints and St Faith, Childerditch, ERO, D/P 230/1/1 (1537–1710).

42. ERO, D/AEW 19/102 (1631), made probate 705 days after writing.

a messuage of six acres with the appurtenances, and twenty acres of freehold land with a barn "late purchased of John Thresher deceassed" (i.e., another John Thresher).[43] Nicholas also received furniture—including a joined bedstead, a cupboard, a small table, a great chest, and a "Clocke in the hall." Another messuage and £40 was left to George Thresher (also a nephew). Additional monetary bequests totaling £84 plus 10s left to each of Thresher's unnumbered household servants complete the will's bequests,[44] and he gave "All the rest of my lands goods Cattells & Chattells vnbequeathed" to his nephew Nicholas. Thresher signed his will.

Nine months after she had buried her husband, Phoebe Thresher had her own will drawn up, which she signed with a mark.[45] In it she bequeathed to her cousin Thomas Veare of Bulphan (three miles from Childerditch):

> one ffeatherbed, one boulster, three blankets & a rugge couering, lying in the Chamber wher my husband dyed. one other ffeatherbed one bolster, three pillowes, three blankettes & a greene Rugge, wth the curtaines and valance, in the Chamber where I lye my selfe. . . . halfe a dozen payre of flaxen sheets one payre of pillow beeres, sixe pewter platters of the middle sort, two small brasse kettles, & two Cowes, the one called Offine, & the other litle bigges. Item I giue & bequeath vnto my cousin Joseph Veare, my Gelding & two Cowes, the one called high-horne, & the other Rugge. Itm I giue and bequeath vnto my cousin Pheobe Reeue two Cowes, the one called blackbird, the other sawen-horne . . .

---

43. Two contemporaries called John Thresher are recorded in the parish register of All Saints and St Faith Church, Childerditch: John, son of Nicholas Thresher, baptized on 19 November 1564, and John, son of Richard Thresher, baptized on 7 September 1569. Two other near-contemporary John Threshers are also recorded in the register of baptisms: one (whose father's name was not recorded) on 13 August 1557 and the other (the son of Richard) on 11 October 1565. No record exists of the death of either of these John Threshers. John "the younger" (as he was called when one of the baptisms was recorded; he is referred to as "junior" in the other) had at least two children: Richard (baptized July 1605) and Ambrose (baptized 14 February 1608). ERO, D/P 230/1/1. The burial of Richard Thresher, son of John, is recorded on 1 January 1612 in the register of St Peter's, South Weald, and Ambrose is mentioned in his mother's 1614 will (ERO, D/AEW 15/52), so it might be assumed that John the testator is John the elder, as he mentions no children. John junior was buried in the church in South Weald in March 1612. Parish Register of St Peter's, South Weald, ERO, D/P 128/1/2 (1559–1654). It is likely that the John mentioned as deceased in John the testator's will was John junior's son, who was baptized in Childerditch in December 1602.

44. Eleven years before Thresher's death, the following entry was recorded in the register of All Saints and St Faith, Childerditch: "Nathan Peachie & Anne Carter servants of John Thresher were marryed on ye vit day of November 1621." ERO, D/P 230/1/1.

45. While there is no record in the Childerditch parish register of John and Phoebe Thresher's marriage, Pheobe Veare, daughter of Thomas, was baptized there on 27 August 1570. Another Phoebe was baptized in the same church in 1567 (family name Piecke), but the fact that Phoebe Thresher's first legatee, her cousin, was called Veare makes it possible that that was also her maiden name. ERO, D/P 230/1/1.

In addition to other bedding, "stayned clothes," household stuff, and a number of monetary bequests totaling over £13, she bequeathed an unnamed "litle yong bullock" to another cousin, Anne Heard, and left "All the rest of my goods ^chattells^ and household stuffe vnbequeathed, my debts payd & my funerall discharged" to her cousin John Veare and to the minister of Childerditch, Daniel Duckfield, who were named as her executors.[46]

Again, there are a number of points of interest in these two documents. First, it is likely that the "Ten Milche Cowes" her husband had bequeathed to Phoebe in his will included some if not all of the cows she gave names to in her own, which was written only two years later: after all, as Leonard Mascall noted, "a Cow will liue well fifteene yeares, but after that she will ware feeble & weary."[47] We might read the fact that John Thresher used the collective term "Cowes" while his wife used individual names as implying that naming animals was something a woman was more liable to do. Phoebe was, after all, likely to be the one who was responsible for the dairy and all that went with it, not her husband, so she would have built up a closer relationship with the dairy herd than he did. As Thomas Tusser put it, "Man cow prouides, / wife dairy guides."[48] The small sample of fifteen wills that include named animals in the large ERO dataset does not reflect this, however: fourteen of them were made by men. Phoebe Thresher is the only female testator in the large ERO dataset who gave her cows names.

Second, the fact that the "Ten Milch Cowes" John bequeathed might include the six cows Phoebe named in her will implies that one does not always need to call a named cow by her name, and this reiterates the possibility that all cows had names even though so few wills reflect this. It also reminds us that even named cows are ultimately just valued property in wills. Indeed, for John Thresher the cows were interchangeable with "ffyftie pounds in money att her Choyse," a calculation that seems to undermine the special status we might assume a named being would have. But—and I return to this—I wonder if it is more complicated than that: that an animal's having a financial value might not discount experiencing grief over it, that the two—money and love—were not as cleanly separated as Anthony Stafford seemed to assume.

Third, the "litle yong bullock" Phoebe bequeathed was not given a name: like Goldelocks's heifer, the new arrival remained anonymous. There are a number of ways to interpret this. It might be that a name was given after

---

46. ERO, D/AEW 19/138 (1632), made probate 28 days after writing.
47. Leonard Mascall, *The Government of Cattell* (London: Tho: Purfoot, 1627), p. 53.
48. Thomas Tusser, *Five Hundred Points of Good Husbandry* (London: I. O., 1638), p. 79.

an animal's character began to emerge: that John Grant of Southchurch, for example, only called his cow Berry once he had seen her eating or trying to push through a hedge.[49] Perhaps the "litle yong bullock" in Thresher's herd was too little, too young to have emerged into his or her full bovine personhood. More likely, however, is the possibility that animals that were marked for selling on or for early slaughter were not given names. This would make sense of the fact that all the bovine animals named in wills were dairy cows. These were the creatures who were long-term residents, partners over an extended period of time. What this interpretation makes possible is the idea that naming signals attachment, that it signals individuality and is not simply labeling: that people named those animals they had, or were going to have, relationships with.

Fourth, like New Come, the names of Thresher's cows were sometimes also descriptions: "litle bigges" is a reference to the size of her teats (one of Josselin's cows was called Smallbigs)[50] and the names "high-horne" and "sawen-horne" seem self-explanatory. This might imply that the descriptions of animals discussed in the previous chapter were more than just descriptions and were in fact names. Indeed, we have already encountered descriptive phrases that are presented as names in wills: "one Called Evered [ever-red] and the other Called the Bullocke New Come,"[51] "one Cowe called the brinded cowe,"[52] and, of course, "a cow called Goldelocks."[53] As these wills show, to be called something might also mean to be it: this is not nominative determinism, but descriptive naming. Of the twenty-seven named cows in the wills, in fact, sixteen might simply be called by their descriptions.[54] In his 1631 will, for example, Richard Taylor, a husbandman from Stanford Rivers, bequeathed "Twoe Cowes one called the branded Cowe and the other the longe legged Cowe" to his son Richard and "Twoe Cowes one called white belly and the other the pyde Bullocke" to his daughter Mary.[55] If we discount the names that are also descriptions, that leaves perhaps as few as eleven cows with what we might call "real" (that is, humanlike) names in the large ERO dataset.

However, fifth (and undercutting that estimation yet again), the cow called "Offine" hints at another method of naming. The use of the human

---

49. ERO, D/APbW 1/15.

50. See *Oxford English Dictionary*, s.v. "bigs."

51. ERO, D/ABW 46/107.

52. ERO, D/ABW 47/212 (1626), made probate 78 days after writing.

53. ERO, D/AMW 3/172 (1617), made probate 46 days after writing.

54. Christine Peters mentions a cow called Fillpayle in a Worcestershire will from the 1570s. See Peters, "Single Women in Early Modern England: Attitudes and Expectations," *Continuity and Change* 12, no. 3 (1997): 325.

55. ERO, D/AEW 19/76 (1631), made probate 144 days after writing.

lineage—the godmother's gift, the father's bequest—to identify an animal has been discussed, and it is possible that the name of the animal might signal something similar. Peter Edwards has noted this in what he terms the "genteel custom of naming a horse after the person who sold it." Such naming, he argues, can help trace upper-class networks.[56] A similar practice might also have existed in relation to cows. One Thomas Offine is recorded in the Parish Register of All Saints and St Faith Church, Childerditch—that "sparsely populated" village—during the time the Threshers were living there. Offine married Elizabeth Ballard in the church on 18 November 1604; he had children baptized there in 1607, 1611, 1614, 1617, and 1624; and he was buried there on 20 March 1627.[57] Perhaps it is possible that the cow "Offine" was named after the man she was bought from (Thomas Offine left no will, so we do not know what his occupation was). If that was the case, the name of the cow would signal—by comparison to the "genteel custom" of naming horses after those who sold them—that for those lower down the social scale, a cow was a valuable possession.

Or maybe it was a neighborly joke to name one's animals after friends. In this context, the name of the bequest by Richard Gosby of "A Red Cowe Called Red hawkins" to his maidservant Ann might reflect not only the animal's color but also what might be termed its human sire.[58] The same might be true for a number of other named cows, for example, Berry, Harris, Sharpe, and Welch—at least one person with the family name of these cows is included in the Essex Record Office's online searchable database during the period 1620–1650 (I have not considered geographical proximity).[59] Indeed, Ralph Josselin's diary seems to reveal something of this kind. In his entry for 27 June 1650, he wrote, "This [day I fetcht] a cow from *Westney's* of Halstead it cost 6li.5s." A year later, when he listed the bulling of his kine, Josselin included his first mention of a cow called "wessney," and on 3 April 1652 he wrote "Wesseny calved. shee tooke but twice; last time was. July. 2. she went not 40. weekes full(.) a blacke bull calfe." "Westney" (i.e., spelled the same way as the family she was bought from) was recorded as having been bulled once again in June 1655.[60] It is possible that this practice of naming the animal after the human sire of a cow was not (or not only) a neighborly joke. Rather, it might be that naming a cow after the person she was bought from

56. Edwards notes that working equines were often given "more prosaic names." Peter Edwards, *Horse and Man in Early Modern England* (London: Continuum, 2007), pp. 15 and 23.

57. ERO, D/P 230/1/1.

58. ERO, D/AEW 18/298 (1629), made probate 94 days after writing.

59. ERO, D/APbW 1/15 lists cows called Berry, Harris, Sharpe, and Welch.

60. Josselin, *The Diary of Ralph Josselin*, pp. 208, 247, 276, 348.

might also work to record for the new owner whether this neighbor provided fertile animals (an early form of what is now termed "progeny testing"), or it might have helped ensure that the bulling of that cow was not done by her own father or brother (Josselin's cows were bulled by a number of different males).[61] This method of naming may, therefore, have had a practical function rather than simply have been done to individualize the animal.

Sixth, like the unfortunate slippage in the phrasing Robert Sabin used when he bequeathed to his wife "the two beasts that she brought [into our marriage] & ye bullock yt was hir calfe,"[62] Phoebe Thresher's bequests of "a rugge couering, . . . a greene Rugge . . . & the other Rugge" make the animal's name ambiguous in a way that is a useful reminder to us of its double status. Given that she seemed to be careful in her naming of pairs of animals— Offine and Little Bigges, Blackbird and Sawen-horne—we can assume that in the other bovine bequest she left "two Cowes, the one called high-horne, & the other Rugge."[63] However the existence of two previous rugs (i.e., coverings)— one written, like "the other Rugge," with a capital R—opens up the possibility that that final "other Rugge" could refer to a third covering. It is more likely, I suggest, that the name is descriptive of this cow's thick rug-like coat, but in the accident of phrasing the name is simultaneously clear and objectifying. As such, Rugge might stand as an exemplar for many of the named cows in the wills: she is an individual and she is also (like) a thing.

Naming practices, it appears, were complicated. As shown in even the small number of examples in the large ERO dataset, they evidence the animal's appearance, relationship to other animals, their (human) source, and their commodity status. But whatever the origin of a name—even if New Come was only New Come because she was the most recent arrival—once it was used, that name became the animal's name. And the description may have been only the formal version of a name. Thus, as the William Walford who was described as "ye younger" in the will might have been "Bill" in day-to-day encounters with his neighbors, so "The litle blacke northerne cowe [who] tooke bull. May. 31 [1651]" in Ralph Josselin's diary was, according to

---

61. See Nicholas Russell, *Like Engend'ring Like: Heredity and Animal Breeding in Early Modern England* (Cambridge: Cambridge University Press, 1986), especially pp. 150–54. Russell makes no mention of an early modern fear of the dangers of incest in breeding practices, so this may not have been a concern at that time.

62. ERO, D/AEW 18/109.

63. The pairing of the animals in the will might have reflected known friendships among them. Henry Buller notes animal welfare scientist Christine Nicol's finding that "farm animals have preferred companions, have complex social behaviour, will spend time in social interaction at the expense of rest time and so on." Buller, "Individuation, the Mass and Farm Animals," *Theory, Culture and Society* 30, no. 7/8 (2013): 167.

his marginal note to that entry, called Norden, something made clear when he noted in March 1652 that "Norden calved a bull calfe shee tooke bull May. 31. 1651. past, which is 40 weekes and 5 dayes."[64] The practices of naming thus reveal that sometimes a name was not just a name, that individualization when it is legible in wills was far from straightforward. But individualization was present and was key to relationships with many animals, and this takes us back into the world of work.

Discussing the academic practice of granting interviewees anonymity so that they might feel able to speak "freely," philosopher and ethologist Vinciane Despret has written, "The result, the stake, even, of anonymity, is to produce a radical asymmetry of expertises." The asymmetry she points to reveals the subjection of the witness to the authority of the "expert."

> The subject is summonsed by a problem that he or she often has nothing to do with, or in any case nothing to do with the manner in which the problem is defined. . . . And most of the time the subject mobilized in this way will agree to respond to questions without calling into question their interest, their appropriateness or even their politeness.[65]

While Despret is here thinking about "scientists' and their "witnesses," her conception of expertise and politeness also includes humans in their relationships with animals, and in an essay titled "Sheep Do Have Opinions," she addresses the failure of ethologists (those who study animals' behavior in their natural environment) to engage with animals in productive ways. She notes that while "primatology has gradually adopted the methods and questions of anthropology[,] classical ethology ... focusses mainly on relations with and around food; who eats what, how animals organize themselves around resources, and so forth."[66] For this reason, primates are enabled by those who study them to reveal things about themselves that other animals are not. But even primatology can get it wrong. As is now recognized, Jane

---

64. Josselin, *The Diary of Ralph Josselin*, pp. 247, 274.

65. Vinciane Despret, "The Becomings of Subjectivity in Animal Worlds," translated by Andrew Goffey, *Subjectivity* 23, no. 1 (2008): 131. Despret is meditating on the granting of anonymity in the context of relegating people she interviewed in refugee camps in the former Yugoslavia to the status of "refugee." This, she feared, actually "prolonged what one could call a regime of insults—'you refugees'" (pp. 130–31).

66. Vinciane Despret, "Sheep Do Have Opinions," translated by Liz Carey-Libbrecht, in *Making Things Public: Atmosphere of Democracy*, edited by Bruno Latour and Peter Weibel (Cambridge MA: MIT Press, 2005), p. 360. Because an error in the original publication has excised part of the text at the bottom of columns, I have also used the complete text at http://orbi.ulg.ac.be//bitstream/2268/135590/1/Sheep%20do%20have%20opinions.pdf, accessed 17 February 2016.

Goodall's team's practice of putting out bananas to habituate chimpanzees in Tanzania to the presence of humans may have had an impact on the chimps' behavior: "provisioning . . . progressively accentuated competition among chimpanzees and produced social disruption."[67] Despret cites as an example of good practice the primatologist Shirley Strum's wish to ensure that "the questions she puts to baboons are always subordinate to 'what counts for them.'" That is, Strum aims never to construct what she calls "'knowledge behind the backs of those I am studying.'"[68]

This is "the virtue of politeness" that Despret argues that all interviewers (whether of humans or of animals) should possess.[69] The virtue enables those being studied or interviewed to display what they think—to show, as her essay's title has it, that they too have opinions. The problem, as Despret sees it, is that those who have studied sheep have asked them the wrong questions, and she turns to analyze the response to this recognition by former primatologist and now sheep farmer Thelma Rowell. Rowell began the practice of putting down an extra bowl when feeding her sheep to avoid "focus on competition, [which was] characteristic of ethological research studies for years." The extra bowl allowed the sheep to "be more interesting": it provided a context in which "if a sheep leaves its bowl, shoves away its neighbor to take its place and immediately returns to its bowl, or persists and follows the other one to oust it once again, a large number of hypotheses can be formulated." Allowing for such behavior by putting down the extra bowl thus expands "the repertoire of possible motives" the sheep are displaying and "allows far more sophisticated explanations." "Of interest," Despret writes, "is he or she who makes someone or something else capable of becoming interesting."[70] The questions an interviewer might frame should be formulated with, and perhaps by, the interviewee, and the extra bowl is Rowell's invitation to the sheep to do just that.

Wanting to interrogate "the possibility that animals could take an active part in work," Despret and sociologist of agriculture Jocelyne Porcher describe the difficulty of asking a question about that issue that the farmers they were interviewing could answer. Their solution was to acknowledge the difficulty, and to—politely—ask the farmers themselves to formulate the

---

67. Interview with Thelma Rowell in Vinciane Despret, "Domesticating Practices: The Case of Arabian Babblers," in *The Routledge Handbook of Human-Animal Studies*, edited by Garry Marvin and Susan McHugh (Abingdon and New York: Routledge, 2014), p. 37.

68. Despret, "Sheep Do Have Opinions," pp. 360–61, quoting Strum, *Primate Encounters*.

69. Despret, "Sheep Do Have Opinions," p. 360.

70. Despret, "Sheep Do Have Opinions," pp. 360, 368, 363.

question. This is how they recall how they invited responses from their inter-viewees. Porcher would say to them:

> "During the research I undertook with farmers I (Jocelyne Porcher) often heard anecdotes, stories, even ways of talking which suggested that ani-mals, in some way, collaborated in work. Now, when I tried to pursue this question with the farmers head on, I was met with resistance or incomprehension. Clearly it's not a good question to ask. But first-hand evidence kept coming up; this encouraged me to persevere. So, in your opinion, as a farmer, how do you think I should be framing my question so that it has a chance of being understood and being interesting?"

The responses of farmers allowed Despret and Porcher to recognize "an amaz-ing similitude between our relationship to research and the relation of the farmers with their animals."[71] The farmers, like the academics, also constantly questioned their own authority. Thus one interviewee, Manuel Calado Varela, noted, "When I open the doors, the cows know I want them to go out, but I don't know if they really want to go out." Animals, he notes, "know what we want, but, we, we don't know what they want." Or, as Despret put it, the cows "pay attention" and are "passably good translators of [human] intentions."[72]

Varela's action—his opening of the doors—was, as he understood it, not performed in a vacuum but in the company and for the attention of respond-ing others. It was a question ("Can you go out here?") that received a polite response from the cow ("Oh, I see. You want me to go out there."). Later work by Porcher and Tiphaine Schmitt studied the animals as much as the farmers and allows us to see that the cows might well know quite specifically what they are being asked to do:

> The only time that Christian [the farmer] usually goes into the stalls is to lay down straw, except when he checks the cows' udders. The cows therefore watch him less in this instance than when he is walking around the housing barn. Once he has gone into the first stall in the row, they know that he is going to carry on until the end because they are familiar with his way of working. The cows are never wrong; they know that this is a time when Christian does not send them to be milked. All they have to do is to get up and temporarily leave their stall.[73]

71. Vinciane Despret and Jocelyne Porcher, "The Pragmatics of Expertise," translated by Stephen Muecke, *Angelaki* 20, no. 2 (2015): 92, 94.

72. Depret, "Becomings of Subjectivity," p. 133.

73. Jocelyne Porcher and Tiphaine Schmitt, "Dairy Cows: Workers in the Shadows?" *Society & Animals* 20, no. 1 (2012): 49.

Despret has noted that such bovine collaboration has always been misunderstood. "When everything happens as it should," she writes, "we don't see the work. . . . When the cows go peacefully to the milking machine, when they do not kick up a fuss . . . we do not see this as testimony to their willingness to do what is expected of them."[74] Rather, "When animals do what they know is expected of them, everything begins to look like a machine that is functioning." Indeed, she turns to philosophy itself and notes: "I would even suggest that what we call mechanistic thought . . . could be partially due to the good will of the animals themselves!"[75] This mechanistic perspective emerges out of the work of René Descartes and posits, as animal welfare scientist Françoise Wemelsfelder puts it, that "animals can evolve to high levels of complexity without needing subjective perspective or experience." She goes on: "Assuming mechanistic causation imposes on discussions of sentience an abstract, technical, mechanical language and rationality that by its very nature removes us from experiential aspects of understanding."[76] Despret's interpretation of animals' good will echoes the difference posited by the early modern law between wild animals who could stray of their "own accord" (and thus had "subjective perspective or experience") and domestic ones who could only "wander" (and so be understood as reacting in a mechanical fashion). And it also allows us to view Descartes as a very poor ethologist: to note that what he saw as revealing animal automatism was, in fact, evidence of the fact that they were better trans-species translators than he was.

It is worth wondering whether translating animals' willingness into a view of their mindlessness is a kind of "calculated forgetting" that takes place because it serves a valuable purpose. In early modern legal terms, eliminating the possibility that a domestic animal could possess a will meant that that animal could be absolute property. Likewise, some modern farming techniques, in their mechanization of animals' lives, their requirement that animals engage in prescribed ways with the processes that are imposed upon them—does not allow those animals to fully present themselves as themselves. That, in turn, reinforces conceptions of animals as capable of reacting but not responding, as lacking the volition that allows a wild animal to walk away and thus to resist ownership. In fact, those who design the processes of industrial farming sometimes do not even acknowledge that the animals

---

74. Vinciane Despret, "From Secret Agents to Interagency," *History and Theory* 52, no. 4 (2013): 42–43.

75. Despret, "From Secret Agents to Interagency," p. 43.

76. Françoise Wemelsfelder, "A Science of Friendly Pigs . . . Carving out a Conceptual Space for Addressing Animals as Sentient Beings," in *Crossing Boundaries: Investigating Human-Animal Relationships*, edited by Lynda Birke and Jo Hockenhull (Leiden and Boston: Brill, 2012), p. 225.

might be able to respond in the first place. Sheep, to use one of Despret's examples, can simply be present and presented as things that, as she puts it, "convert plants into mutton."[77]

Historiography, like industrial farming (but without the appalling death toll), has also too often failed to recognize animals' actions as responses and has rarely asked them polite questions. Farm animals in particular—creatures Henry Buller has termed "orphans not only of ethology but also of history. Perhaps of sociology, anthropology and geography too"[78]—are treated as populations rather than individuals: historians have tended to count them rather than contemplate what it was like to live with them on a one-to-one basis. Because of this, analyses too often close down the potential of the non-human members of early modern society to be present as active, engaged participants. The assumption is, perhaps, that while the husbandman labors, the cow simply ruminates and that such rumination should be the focus of biological, not historical, inquiry. Thus, just as the early modern law regarded anonymity as the norm and utility as the requirement for an animal to have the status of an ownable thing, so the assumption of anonymity in the context of historical assessments of animal agriculture perhaps enables the belief that animals do not work. In this context, just as Despret challenged the anonymizing of the refugees she was interviewing because she feared that it repeated "the process of exclusion" that had led to their being refugees in the first place,[79] so it might be possible to read naming a cow in a will as allowing us to see how the people who worked with them in early modern Essex included them in their worlds and recognized them as co-workers, or "Partners," as Dekker's North-Country-man put it.

So, what would it mean to recognize an animal as a co-worker? This is not a question of terminology, of arguing for politically correct language to enter agricultural history. Rather, it is a question of taking seriously that the yard and the field are shared spaces, that opening a door for a cow is an action that assumes a response: that it is, as Vicki Hearne said of dog training, an

---

77. Despret, "Sheep Do Have Opinions," p. 362. The work of Temple Grandin to redesign slaughterhouses in the light of her understanding of bovine experience runs counter to the tendency to treat animals as parts of the larger machine. In addition, Lewis Holloway and colleagues' study of how robotic milking machines have allowed cows "freedom to choose when and how often to be milked" signals this as well. See Temple Grandin and Catherine Johnson, *Animals in Translation: The Woman Who Thinks Like a Cow* (London: Bloomsbury, 2005); and Lewis Holloway, Christopher Bear, and Katy Wilkinson, "Re-Capturing Bovine Life: Robot-Cow Relationships, Freedom and Control in Dairy Farming," *Journal of Rural Studies* 33 (January 2014): 131–40, quote from Holloway et all, on p. 134.

78. Buller, "Individuation, the Mass and Farm Animals," p. 161.

79. Despret, "Becomings of Subjectivity," p. 130.

invitation to an animal to participate in a conversation.[80] In this context, a rug is not for lying on but is for negotiating with, for talking to, because a successful workplace requires decent lines of communication; it runs most smoothly when there is a shared understanding of activities. And this shared understanding would have been even more likely back then because in early modern agriculture, small numbers of cows were often tended by the same few people and milking took place by hand.[81]

What kind of conversations might have taken place? Inevitably, they would have been routine: Gervase Markham's description of the husbandman's day spells out the repetitive nature of the chores with animals—cleaning, feeding, feeding, cleaning—so conversations would have frequently been formulaic. This was and is something that cows would have liked, as Ralph Josselin knew when he noted that he "had great trouble in the cowes unquietnes" when he brought Wessney home from Halstead (a journey of about three miles).[82] Familiarity is reassuring to cows and this animal was upset by the strangeness of the experience of relocating.[83] In addition to the routine, however, it is possible that in the fields and yards of early modern England, there may have been variety in the human-cow conversations as well. To be involved in a conversation is, after all, sometimes, to hear things you do not expect or want to hear and to have to respond to such unwished-for responses. Indeed, writing of her and Porcher's work with farmers, for example, Despret notes that "we were demanding the maximum of kindness in relation to our research, while creating the conditions for maximum 'recalcitrance.'"[84] They were constantly inviting the farmers to point out the problems with their mode of address. The same might be true of early modern housewives and their cows. Inviting kindness and allowing for recalcitrance is putting down the extra bowl, is acknowledging the expertise of the interviewee. On the most basic level, it

---

80. See Vicki Hearne, *Adam's Task: Calling Animals by Name* (Pleasantville: Akadine Press, 2000), pp. 42–76.

81. The first milking machines began to appear in the 1830s, and the "real beginning of modern milking," according to G. E. Fussell, came with the "pulsator principle," or the machine's replication of a calf's sucking or hand milking, in 1895. Fussell, "The Evolution of Farm Dairy Machinery in England," *Agricultural History* 37, no. 4 (1963): 223. It is interesting to note that modern farmers who use automated milking machines that allow the cows to choose the time to be milked feel that this apparent distance from the animals (i.e., the fact that the farmers are not actively engaged in herding them into the milking parlor, for example) actually increases intimacy. Not only are the cows allowed to choose when to be milked, which means happier animals, but the farmers are also freed up from herding and can spend more time checking on and being with them. See Holloway, Bear, and Wilkinson, "Re-Capturing Bovine Life."

82. Josselin, *The Diary of Ralph Josselin*, p. 208.

83. On cows' preference for routine over new experiences, see Grandin, *Animals in Translation*.

84. Despret, "Becomings of Subjectivity," p. 132.

was understanding which side a cow liked to be milked on and going along with that; it was thus allowing the animal to occasionally dictate the direction of the exchange.

As with so many of the day-to-day experiences of early modern livestock, these polite encounters between humans and cows are almost wholly lost to us. No sonnets, no treatises, no epics have been written that preserve these conversations, yet we get glimpses that allow us to see that they did take place. The milkmaid is often our source. Like her cows, she was written about rather than writing, but the advice, the stories, the comic anecdotes about her reveal something of their entangled lives. Their closeness is made clear in the issue of recognition. Cows knew their milkers and required—you might say demanded—that the same person approach them every day. Bartholomew Dowe's "Suffolcke man" recalls that on his large dairy farm "I haue had also this experience, that one of my kine hath had such a minde and fantastie to one of my Maides, that in her presence the Cow would neuer stand to be milked of anie other but of her onelie." But he goes further than this: the cows' fear of the new can be traced in their demand that the maids wear certain clothes: "when Maides come to milke kine more gailiar apparelled, then they be accustomed to milke in, the kine will be verie dangerous to stande to bee milked of them."[85] Here, the door is being firmly closed—by the cows. But cows could also open the door to conversation: Markham wrote that the woman "shall not settle her selfe to milke, nor fixe her paile firme to the ground till she see the Cow stand sure and firme."[86] Gaining the cow's agreement was vital to success. The routine of milking might also include music: as Dowe noted, "if your milking Maides be disposed to sing in time of their milking, some Cowe will take such a delight therein, that afterward whē a Maide commeth to milke her and doth not sing, shee will not stand to be milked."[87] Richard Brathwaite, in a critical assessment of "ignominious" balladmongers, claimed that "every poor Milk maid can chant and chirpe [a ballad] under her Cow; which she useth as an harmlesse charme to make her let downe her milke."[88] However, music historian Christopher Marsh has noted that such singing might not be so "harmlesse": "milkmaids," he writes, "knew by instinct and experiment that singing was good for milk yields."[89]

---

85. Bartholomew Dowe, *A dairie booke for all good huswiues* (London: Thomas Hacket, 1588), C1v.

86. Gervase Markham, *The English Hovse-Wife* (London: Anne Griffin, 1637), p. 194.

87. Dowe, *A dairie booke for all good huswiues*, C1v.

88. Richard Brathwaite, *Whimzies: Or, a new cast of characters* (London: Felix Kingston, 1631), pp. 10, 12.

89. Christopher Marsh, *Music and Society in Early Modern England* (2010; repr., Cambridge: Cambridge University Press, 2013), p. 39. In 2001, an experiment by researchers at Leicester University

Gentleness, as well as routine, was key to successful milking: Dowe noted that cows who needed to be brought "farre to be milked at home" should "bee quietly brought" as otherwise "you shall haue thereby much the lesse milke": haste would make waste with these animals. He also recognized how important it was for cows to be "well milked and stroked," noting that the alternative would be "marring to the kine, for thereby they will the rather grow drie, and be the worse milch long time after."[90] This could be an example of the "anti-cruelty ethic" of the social contract between humans and livestock Bernard E. Rollin has identified.[91] Here gentle treatment is seen to lead to high milk production and thus benefits both humans and cows, while harsh treatment—the cows being "hastilie dryuen"[92]—leads to low productivity. A similarly self-serving reason for kindness can be traced in Thomas Fuller's bovine metaphor in his *Historie of the Holy Warre*. He used it to explain what he viewed as the cleverness of Henry III in saving many Englishmen from foolish involvement in the Crusades: "The King promised he would go with them, and hereupon got a masse of money from them for this journey. Some say, that he never intended it; and that this onely was a trick to stroke the skittish cow to get down her milk."[93] The image seems to reveal the king's cunning in using the opportunity to squeeze cash out of his subjects, but there is an alternative reading that presents him in a better light. If the metaphor is translated back to its original source, a skittish cow that gave little or no milk would be a useless animal and would be heading for the slaughterhouse, so stroking her would encourage milk production and would give her a value and potentially save her from death. It is telling—and perhaps accidental—that in this metaphor, milking is presented as working both for and against the cow. It might be an image about tricking someone out of something, but

---

revealed that "on average, each cow produced 0.73 litres more milk per day" when listening to slow music (less than 100 beats per minutes) than when they listened to fast (over 120 beats). See Melanie Cooper, "Milking the Music," *New Scientist*, 30 June 2001, p. 12. It is not just dairy cows who like music; Henry Chettle wrote that "the Carman [carter] . . . was woonte to whistle to his beastes a comfortable note." Chettle, *Kind-Harts Dreame* (London: William Wright, 1593), C3v.

90. Dowe, *A dairie booke for all good huswiues*, B3v.

91. Bernard E. Rollin, *Farm Animal Welfare: Social, Bioethical, and Research Issues* (Ames: Iowa State University Press, 1995), pp. 5–6, 8.

92. Dowe, *A dairie booke for all good huswiues*, B3v.

93. Thomas Fuller, *The Historie of the Holy Warre* (Cambridge: John Witham, 1639), p. 198. I found this text through a search using the phrase "skittish cow" on the Early English Books website (http://corpus.byu.edu/eebo/, accessed 21 July 2017). I first encountered the phrase in the ballad *The woman to the Plow and the man to the hen-roost; or, A fine way to cure a cot-quean* by M. P. [Martin Parker] (London: F. Grove, 1629). The only other usage of the phrase recorded in the byu.edu corpus, in William Vaughan's *The golden fleece* (1626), also uses the phrase to describe human rather than bovid actions.

it might also be read as an analogy of salvation. The cow, it seems, could be regarded as both victim and agent in the dairy.

The danger of the skittish cow to herself is reiterated in Markham's discussion of the politeness expected from the cows as well as the humans in the partnership. He gave this a particular meaning:

> Touching the gentlenesse of kine, it is a vertue as fit to be expected as any other; for if she be not affable to the maide, gentle and willing to come to the paile, and patient to have her dugs drawne without skittishnesse, striking or wildnesse, she is utterly unfit for the dayry.[94]

Not only does Markham's description seem to allow for bovine "wildnesse" (something the law denied), he also figures the cow as being affable, from the Latin *affabilis*, meaning approachable. It is notable that when the OED defines the term as "of a person, or a person's character or bearing: easy to approach and converse with," the possibility of animals' approachability is forgotten and, yet again, the potential for their collaboration is effaced.[95] Whether Markham is anthropomorphizing when he uses the term to describe a cow's behavior, therefore, or is simply describing the bovine character necessary for a dairy cow is unclear, but his choice of the term "affable" signals his understanding that the relationship is two way; that the conversation of milking is made up of questions and responses. In a similarly reciprocal vein, Nicholas Breton noted that in "that goodly *golden time*" when dogs were dogs, "*maidens* sate and neately milkt their *Cowes*."[96] "Neately" here does not mean "tidily" but "skillfully; cleverly, dextrously."[97] The fact that Breton presents this kind of milking as part of a lost golden time signals that something had changed by the early seventeenth century. Perhaps growing herd sizes were making neat milking less likely. The shift might have been happening at the time Breton was writing: in his 1588 work, Dowe has his Suffolk farmer state that one dairy maid on his farm milks twenty cows whereas his female interlocutor from Hampshire suggests that "eight or nine kine is enough for one maide seruaunt to milke in this county."[98]

---

94. Markham, *English Hovse-Wife*, p. 192. During research, I chanced upon the will of John Lane, a gentleman of Twickenham in Middlesex, who bequeathed to his daughter Margarett "one Blacke Cowe called Gentle" in his 1597 will. NA, PROB 11/89/282 (1597), made probate 379 days after writing.

95. *Oxford English Dictionary*, s.v. "affable."

96. Breton, *Old Mad-cappes new Gally-mawfrey*, D2v.

97. *Oxford English Dictionary*, s.v. "neatly."

98. Dowe, *A dairie booke for all good huswiues*, A3r. Using data from Norfolk, Bruce M. S. Campbell and Mark Overton trace the "doubling of stocking densities" in the seventeenth century, seeing this as what "most distinguishes early modern from medieval agriculture." Campbell and Overton, "A New

But there is more to the word "neatly" in Breton's usage than a sense of nostalgia for a lost intimacy between milkmaid and cow. In this period, in addition to meaning tidily and skillfully, "neat" was also a term for a bovine animal.[99] Indeed, in *The Winter's Tale*, Shakespeare has Leontes, the troubled monarch, note the homonym: "We must be neat," he says, "not neat, but cleanly, captain: / And yet the steer, the heifer, and the calf / Are all call'd neat."[100] Breton, like Shakespeare later, saw the pun and used it, but in another way: the maids who sat and "neatly milkt their *Cowes*" were, perhaps, exercising true affability: they were taking the milk like calves.

This becoming-calf might seem far-fetched, but comprehension of the cow-calf relation not only recognized that nature offered the best example of how to take the milk, with the action of milking mimicking the calf's sucking motion (as also happened in the nineteenth century in the design of milking machines),[101] early modern writers also envisioned an emotional bond between the cow and her offspring that was a requirement for the successful dairy and was thoroughly recognizable. Thus Markham wrote in his address to the English housewife:

As a Cow must be gentle to her milker, so she must be kind in her own nature; that is, apt to conceive, & bring forth, fruitfull to nourish, and loving to that which springs from her; for so she bringeth forth a double profit; the one for the time present which is in the Dairy; the other for the time to come; which is in the maintenance of the stocke, and upholding of breed.[102]

For Markham, maternal instinct and money slide into one another and being a good mother is regarded as necessary to being a profit-making thing. And so a paradox emerges: cows are like human mothers (or maybe they are just mothers: this too may not be anthropomorphism). But they are also objects for human use. Anthropocentrism, it seems, tops any acknowledgment of similarity.

Shakespeare recognized the potential meaning to be made from bovine animals and in doing so reiterated the perceived correspondence of human

---

Perspective on Medieval and Early Modern Agriculture: Six Centuries of Norfolk Farming c. 1250–c. 1850," *Past and Present* 141 (November 1993): 41, 90.

99. *Oxford English Dictionary*, s.v. "neat."

100. William Shakespeare, *The Winter's Tale*, in *William Shakespeare: The Complete Works*, edited by Stanley Wells and Gary Taylor (Oxford: Clarendon, 1988), 1.2.125–27.

101. Fussell, "Evolution of Farm Dairy Machinery," pp. 223–24.

102. Markham, *English Hovse-Wife*, p. 193. It is interesting that Markham notes the same dual usage of cows in *Cheape and Good Hvsbandry* (a text with an implied male reader), but the question of bovine emotion is absent: "The vse of the Cow is twofold, either for the Darie, or for breed" is all Markham writes. Markham, *Cheape and Good*, p. 88.

and cow in the period. In 3.1 of *The First Part of the Contention of the Two Famous Houses of York and Lancaster*, as the Duke of Gloucester, one of King Henry's key advisors, is removed from his side by his argumentative queen and her followers, the weak king is left to mourn. "Thou never didst them wrong, nor no man wrong," he says of Gloucester:

> And as the butcher takes away the calf,
> And binds the wretch, and beats it when it strays,
> Bearing it to the bloody slaughterhouse,
> Even so remorseless have they borne him hence;
> And as the dam runs lowing up and down,
> Looking the way her harmless young one went,
> And can do nought but wail her darling's loss;
> Even so myself bewails good Gloucester's case
> With sad unhelpful tears, and with dimmed eyes
> Look after him, and cannot do him good,
> So mighty are his vowèd enemies.[103]

Here the king regards the tie of the cow to her calf as unquestionably, and recognizably, emotional. Indeed, the analogy only works because Shakespeare could assume that his theater audience would also share this understanding (we are not, I think, meant to view the king as a fool here). And yet the cow's emotional response—as the responses of Smallbigs, Wessney, Redbacke, Brownbacke and Norden would have been when Josselin removed their calves—is "unhelpful." And we can take this word to signal, I suggest, both that the cow's response is ineffective (it doesn't change anything), and that it is an annoyance—it unsettles the work of the dairy by showing that cows have opinions too. It is, perhaps, one of the most troubling, but also expected, kinds of recalcitrance the human-dairy cow relationship has to allow for.

And here is what for us is the crucial contradiction in envisioning working with animals in early modern agriculture as collaboration. Where modern, intensive, industrial farming might allow for "distancing manoeuvres," for the

---

103. Shakespeare, *The First Part of the Contention of the Two Famous Houses of York and Lancaster*, 3.1.209–20. On Shakespeare's representations of cows and calves see Katherine Duncan-Jones, "Did the Boy Shakespeare Kill Calves," *The Review of English Studies*, New Series, 55 no.219 (2004), 183–95. Dutch playwright Joos Klaerbout dwelled on human separation from a calf in *De klucht van 't kalf* (1662), a translation from a French original. On selling the animal, the farmer states: "Adieu little calf, I won't feed you sweet milk any longer; / My sweet and cherished little calf, this will cost your young life." Quoted in Johan Koppenol, "Noah's Ark Disembarked in Holland: Animals in Dutch Poetry, 1550–1700," in *Early Modern Zoology: The Construction of Animals in Science, Literature and the Visual Arts*, edited by Karl A. E. Enenkel and Paul J. Smith (Leiden: Brill, 2007), 2:470.

depersonalization of animals because scale is so overwhelming,[104] the same shifts seem more difficult when numbers are much smaller; when all cows might have names.[105] What is clear, however, is that finally, unavoidably, early modern animal agriculture saw animals as things for use; it was a system in which conversations always existed within a fixed hierarchy in which the needs of the humans ultimately overrode those of animals. On the enclosed farms and smallholdings of early seventeenth-century Essex, even the named cow was an instrument of supply: if she ceased to supply—if she stopped being fertile or her milk dried up—she would be sent for slaughter. Remember Mary Archer of Little Clacton's testamentary request that "I will that myne executor tak in my two old cowes at Micholmas that are now in the possession of Henrie Soyars and that he shall sell them and putt out two young Cowes in sted of them to the vse and benefitt of John Archer and Robbecah Archer the two youngest children of my sonne Thomas Archer."[106] The sale would have been to the butcher: such pragmatism was fundamental.

The love of a cow for her calf has a paradoxical place in early modern thinking that offers some insight into the wider implications of the individualization of farm animals, then. The love was recognized and humanized the cow, or animalized the human, or simply asserted that humans and cows were not so distinct after all (the ambiguity here is significant, I think). And it was also regarded as good for making a profit—a "good" that positions the cow as an object for human use. But maybe it is even more morally uncomfortable than that "and" reveals. Perhaps the profit-making relied on the existence of the cow's emotional attachment to her calf.[107] Perhaps her personhood—evidenced in her similarity to the humans she worked with, in their mutual ability to communicate and collaborate—was central to a successful household. Such a reading challenges Tim Ingold's sense that for

---

104. Lindsay Hamilton and Nik Taylor, *Animals at Work: Identity, Politics and Culture in Work with Animals* (Leiden and Boston: Brill, 2013), p. 20.

105. Studies of contemporary farmers and their attitudes toward their livestock reveal that closeness persists, but perhaps not for all animals. Using interviews with farmers as her source, Rhoda M. Wilkie suggests that cows' distinct responses to farmers led to them having different relationships with them. These might range from "emotional affinities to and aloofness from their animals." Circumstances also could impact human-cow relations: a cow who had once required "additional attention" changed how that animal was subsequently viewed, as did the animal's length of stay on the farm. Wilkie, *Livestock/Deadstock: Working with Farm Animals from Birth to Slaughter* (Philadelphia: Temple University Press, 2010), pp. 130–31. I return to contemporary ideas in the afterword.

106. ERO, D/ABW 43/197 (1620), made probate 107 days after writing.

107. There is no mention in Mascall, Markham, Heresbach, or Tusser of the possibility of a cow rejecting her calf, and thus the relationship seems, by implication, to be simultaneously recognizable, necessary, expected, and instinctive.

pastoralists "the animals are presumed to lack the capacity to reciprocate."[108] It proposes, indeed, that individualization is a part of pastoral relations and does not reduce the animal's utility but is, rather, central to it. To know an animal by name is not simply to reveal a distinction between the law and the yard, neither is it to show how neighborliness might play a role in successful farming (neighborliness that includes animal as well as human neighbors). It is, in fact, to have found the most efficient means of using that animal. Friendliness means fertility; partnership means profit. They are our nurses; we are their calves. And their butchers.

---

108. Tim Ingold, *The Perception of the Environment: Essays on Livelihood, Dwelling and Skill* (2000; repr., London and New York: Routledge, 2011), p. 72.

CHAPTER 4

Other Worldly Matter

The Immaterial Value of Quick Cattle

The Essex clergyman Arthur Dent began his extraordinarily popular book *The plaine man's path-way to heauen* with a shift that offers an interesting assessment of the place of husbandry in early modern England. Dent's dialogue, which was first published in 1601 and had gone through twenty-five editions by 1640,[1] opens with Theologus, "a Diuine," and Philagathus, "an honest man," out walking at one o'clock one May afternoon. They see approaching them two "neighbours of the next Parish" and decide to engage Asunetus, "a very ignorant man," and Antilegon, "a notable Atheist, and cauiller against all goodnesse," in conversation in order "to speake of matters of religion." They discover that the two men have come to talk to one of Theologus's parishioners about a cow he has for sale. Antilegon notes: "I am afraid at this time of the yeare wee shal finde deare ware of her." "How deare? What do you think a very good Cow may be worth?" Theologus asks. "A good Cowe indeed at this time of the yeare is worth very neere foure pound, which is a great price," Antilegon answers, and Theologus assents: "It is a great price in deed."[2]

---

1. See Brett Usher, "Dent, Arthur (1552/3–1603)," in *Oxford Dictionary of National Biography* (Oxford: Oxford University Press, 2004).

2. Arthur Dent, *The plaine man's path-way to heauen* (London: Melchiside Bradwood, 1607), pp. 1–3. It is worth noting that the cost of the cow did not increase over the twenty-five editions published in the period 1601 to 1640.

At this point Philagathus, the honest man (his name means "lover of the good"), interrupts their conversation: "I pray you M. *Theologus*," he states, "leaue off this talking of kine, and worldly matters, and let vs enter into some speech of matters of religion, whereby we may do good, and take good one of another."[3] For Philagathus, cows are material creatures—"worldly matters"—and in his opinion, focusing attention on them has the potential to drag people down into the mire of the flesh. It is, he urges, the better part of humanity to remember this and "take good one of another" by contemplating otherworldly issues.

In distinguishing cows from spiritual concerns, Dent's honest man echoes the advice of those who urged their readers to write their wills in good health so that their final thoughts might be turned toward the eternal: advice that was, as we have seen, internalized by one Essex testator, who stated that he was "desirous not to be troubled or encombred when it shall please god to visite me wth sicknes."[4] But in assuming that an interest in cows was a distraction from the realm of the eternal, Philagathus may have missed a crucial aspect of human-livestock relations. It might be supposed, as he does, that for the will writers of early modern Essex, quick cattle were simply for use, even as conversations with them were believed to be possible. It might appear unproblematic to imagine that a cow in a will was always just a cow, even if she had a name. The previous chapters would seem to support this reading, with the lived experiences of the field and the yard seemingly as distant from figurative meaning as they were from some aspects of the law. But evidence from the large ERO dataset suggests that the animals people lived with might also have carried a value that went beyond fertility, labor, or financial worth.

This chapter will trace this idea by going on a winding journey. Beginning with a focus on bees, the smallest creatures to be found in any will in the dataset, it will then turn to think about lambs, before contemplating how animals' deaths fit into the domestic economy that underpins so many Essex wills. The journey is guided by a single idea, even if it is not being undertaken alongside a single creature: that the realms of the worldly and the otherworldly were not always as distinct as Philagathus proposed, that sometimes the animals people worked with were meaningful in more than material terms. In this chapter, the routines of husbandry that can be traced in wills are read in conjunction with what might appear to be very different kinds of representation to reveal something else about daily life in the early

---

3. Dent, *The plaine man's path-way*, p. 3.
4. ERO D/AMW 1/71 (1625), made probate within 195 days of writing.

seventeenth century. So, bequests of bees are analyzed alongside contemporary beekeeping manuals that reveal bees to be emblematically as well as practically important; bequests of lambs are read alongside religious works that suggest the possibility that the bequests had spiritual as well financial value; and, finally, the killing of animals—that inevitable conclusion to the relationship people had with so many of them—is addressed through a reading of practical, imaginative, and religious works in which slaughter is revealed to have meaning that goes beyond material utility. The wills have a different role in this chapter. They are not used to lead the way in a reconstruction of early modern life alongside animals; rather, particular bequests are read as reflecting a wider set of ideas that might seem to exist outside of the world of husbandry in the realm of symbolism.

To begin this undertaking, we need to leave the yards and fields of early modern Essex and turn briefly to books—not legal texts this time, but works of what was termed natural philosophy. Here we can see what appears to be a very different vision of the animals with whom people lived. This chapter will argue, however, that this is a vision that was actually shared by those who wrote with animals and those who worked with them.

What William B. Ashworth Jr has termed the "emblematic worldview" was central to a key, literate assessment of animals in the sixteenth and early seventeenth centuries.[5] This perspective did not offer "anatomical description, or [discussion of animals'] place in a taxonomic scheme based on physical characteristics," nor did it read animals as creatures with economic value. Instead, Ashworth notes, natural philosophy regarded them as "living characters in the language of the Creator" that might reveal truths about the Almighty and his creation.[6] In this tradition, the Bible was regarded as the first book of God and the natural world the second, and both were to be studied by the pious individual. Thus, in the early seventeenth century, Edward Topsell proposed that his study of birds, *The Fowles of Heaven*, "teacheth the meaninge of textes of Scripture."[7] This was Topsell's third work on the natural world, following his *Historie of Foure-Footed Beastes* (printed in 1607) and *History of Serpents* (1608).

---

5. William B. Ashworth Jr., "Emblematic Natural History of the Renaissance," in *Cultures of Natural History*, edited by N. Jardine, J. A. Secord and E. Spary (Cambridge: Cambridge University Press, 1996), p. 30.

6. William B. Ashworth Jr., "Natural History and the Emblematic Worldview," in *Reappraisals of the Scientific Revolution*, edited by David C. Lindberg and Robert S. Westman (Cambridge: Cambridge University Press, 1990), pp. 306, 308.

7. Edward Topsell, *The Fowles of Heauen or History of Birdes* (ca. 1613–14), edited by Thomas P. Harrison and F. F. David Hoeniger (Austin: University of Texas, 1972), p. 17. This book remained unfinished at his death in 1625.

His task in these three texts was to bring together the various references to each creature that could be found in religious and classical sources, as well as in more recent material and from these draw out the animal's multiple meanings. As a result of this gathering of information, Topsell believed, the work of God would become more visible and thus more meaningful. In the chapter on "The Alcatraz" (i.e., the pelican) in *The Fowles of Heaven*, for example, reports of the selfless parenting of this bird—it was said that it would run into fire to save its young—allowed Topsell to draw a comparison between it and "Our Sauiour Christ," and the natural and the supernatural were linked with animals revealed, in Ashworth's words, as "just one aspect of the intricate language of metaphor, symbols and emblems" through which divine will might be better understood.[8] A close reading of God's creatures was believed to disclose meaning that might inform humans about the most righteous way of living.

The emblematic readings of animals thus focused on their revelatory and exemplary potential as well as, and sometimes instead of, more worldly aspects of their being or well-being, with books rather than the animals themselves as the key source for authors. As such, a natural philosopher did not need to have encountered an animal to be able to offer an assessment of it, because the assessment emerged from his reading more than from his lived experience. Indeed, animals that would never be encountered, such as unicorns and manticores, were present in these texts because, as Peter Harrison notes, "the criterion for inclusion . . . was not whether something existed in the world, but whether it existed in books."[9] The emblematic worldview would seem, therefore, to be very distant from the sphere of the husbandman who labored for hours every day with his plow team and of the housewife who leaned up against the flank of her cow, morning and evening. But this was not the case. In opposition to Philagathus's conception of the division of cows from matters of the spirit,[10] thinkers such as Topsell believed that it was through animals that people who could not read might come to encounter the divine. Works such as his were, as Gordon L. Miller puts it, to be "recommended to . . . readers more as a guide to soulful meditation than as a guide to the

---

8. Topsell, *Fowles of Heauen*, pp. 36–37; Ashworth, "Natural History and the Emblematic Worldview," p. 305.

9. Peter Harrison, *The Bible, Protestantism, and the Rise of Natural Science* (Cambridge: Cambridge University Press, 1998), p. 66.

10. It is interesting to note that the tradition of natural philosophy, which was declining in importance in the early seventeenth century, is not reflected in other works that wholly condemn the material realm as a distraction. For example, in Dent's work, Theologus feared that the human who thought "of nothing else day and night, but this world, this world, lands and leases, grounds and liuings, kine, and sheepe, and how to wax rich" would only perform actions that would "most manifestly declare, that they are of the earth, and speake of the earth." Dent, *The plaine man's path-way*, p. 19.

field."[11] And Harrison has argued that the promotion of "the virtues of the book of nature provided a theology for the unlettered."[12]

Thus it is no coincidence that Topsell the natural philosopher was also a clergyman. On a practical level, in a world of constant and close interaction between people and their animals, a preacher's understanding of those animals might be useful because it could provide grounds for social interaction—this could be why Dent's Theologus was so engaged in the conversation with Antilegon about cows before Philagathus interrupted them.[13] But knowing the value of animals might go beyond having an ability to converse with one's parishioners about day-to-day matters of importance to them, and in *A Priest to the Temple*, for example, George Herbert advised that a good country parson "condescends even to the knowledge of tillage, and pastorage, and makes great use of them in teaching, because people by what they understand, are best led to what they understand not."[14] Images from husbandry, in short, might be used in sermonizing to encourage people to engage more readily in religious contemplation. But it is not only in sermons that we can trace the emblematic meaning of husbandry. Bequests in the large ERO dataset also reveal that the symbolic potential of animals was something that might be found in fields and yards of Essex. And just as some testators internalized the need to write a will in advance of illness in order to avoid having to think about the material realm on their deathbed, so others' wills seemed to reflect the idea that their animals were "living characters in the language of the Creator," that they had more than just worldly meaning. Bees, I suggest, offer a useful starting point here.

Bequests of bees are rare in the large ERO dataset: only seven wills include them.[15] This could be for a number of reasons. Hives—or skeps, as they were

---

11. Gordon L. Miller, "The Fowls of Heaven and the Fate of the Earth: Assessing the Early Modern Revolution in Natural History," *Worldviews: Environment, Culture, Religion* 9, no. 1 (2005): 68.

12. Harrison, *The Bible, Protestantism, and the Rise of Natural Science*, p. 197. Vibeke Roggen makes a similar point about Wolfgang Franzius's 1612 *Historia animalium sacra*, which was, she argues, "meant for theologians, meant for priests, meant for teaching." Roggen, "Biology and Theology in Franzius's Historia Animalium Sacra (1612)," in *Early Modern Zoology: The Construction of Animals in Science, Literature and the Visual Arts*, edited by Karl A. E. Enenkel and Paul J. Smith (Leiden: Brill, 2007), p. 129.

13. In fact, Ralph Josselin's diary shows that preachers were often themselves involved in animal agriculture. The 1626 will of Edmund Turner, the minister of Chappel, with its gendered bequests of "i cowe" to daughter Margaret, "i cowe" to daughter Jane, and "all my bookes" to son Edmund also reveals this. ERO, D/ACW 10/148 (1626), made probate 378 days after writing. Turner also bequeathed "butter cheese and my hogges" to his wife and daughters. Only three clergymen's wills appear in the large ERO dataset and Turner was the only one to include a specific animal bequest.

14. George Herbert, *A Priest to the Temple, or, The Countrey Parson his Character, and Rule of Holy Life* (London: T. Maxey, 1652), p. 10. The text was published nineteen years after Herbert's death in 1633.

15. This infrequency is not out of kilter with Francis W. Steer's findings. His research showed that in the twenty-two Essex inventories dating from 1635 to 1640, only one included bees, whereas

frequently called in this period—would have been likely to have been attached to or in the garden of the house of the beekeeper and so might have passed as appurtenances in the bequest of the property.[16] But the status of bees is interesting as well. They were not listed in William Lambarde's discussion of stealable animals (quoted in the previous chapter), and such a legal representation may have discouraged testators from specifying them in bequests. But their position was ambiguous in other ways too that might have had influenced perceptions of them as not-quite-ownable creatures. In his 1601 translation of Pliny's *Historia naturalis*, for example, Philemon Holland called bees *"halfe tame,"*[17] and in his husbandry manual, Conrad Heresbach wrote that bees "cannot be rightly tearmed either wilde, or tame."[18] Control (and thus ownership) of these creatures was felt to be tentative. But the presence of bees in seven wills signals that these animals were considered to be ownable, if only by a few testators. Indeed, in his translation of Charles Estienne's *Maison Rustique*, Richard Surflet termed them "this poore cattell," thus placing them within that category—their poorness being a reflection of their size rather than, say, their being put upon.[19]

In their study of the bequeathing of hives in early modern England, Penelope Walker and Eva Crane noted that "most beneficiaries [of bees] were members of the family,"[20] and the bequests in the large ERO dataset reflect this, with one anomaly. John Fellgate, a yeoman from Ramsey, is the only testator to leave bees to someone outside his family; to John Jefre, his local minister. Or, rather, Fellgate offered Jefre a choice of ten shillings "or else a skepe of my beese which he doth beste excepte vpon condicion he shall macke a sarman at my fenerall."[21] Including a legacy for a funeral sermon was not common but did occur in other wills in the large ERO dataset. The bequest

---

fourteen included other animals. Francis W. Steer, ed., *Farm and Cottage Inventories of Mid-Essex 1635–1749* (Colchester: Wiles and Son for Essex County Council, 1950), pp. 71–86.

16. The design and siting of hives, or skeps, in the period is outlined in Arthur MacGregor, *Animal Encounters: Human and Animal Interaction in Britain from the Norman Conquest to World War One* (London: Reaktion, 2012), pp. 394–99.

17. Pliny, *The historie of the world: Commonly called, the naturall historie of C. Plinius Secundus*, translated by Philemon Holland (London: Adam Islip, 1601), 7:242. Other animals that Pliny classed as "halfe tame" were hares, swallows, and dolphins.

18. Conrad Heresbach, *Foure bookes of husbandry . . . Newely Englished, and Increased, by Barnabe Googe* (London:Richard Watkins, 1577), 168r.

19. Charles Estienne, *Maison Rustique, or The Countrey Farme* (London: Adam Islip, 1616), p. 328. For poor as meaning "insignificant" see *Oxford English Dictionary*, s.v. "poor," definition 2d.

20. Penelope Walker and Eva Crane, "English Beekeeping from c. 1200 to 1850: Evidence from Local Records," *The Local Historian* 31, no. 1 (2001): 8.

21. ERO, D/ACW 11/102 (1630). Among the six testators that include bequests of bees, two bequeathed them to sons, one to a wife, one to a sister, and one to a grandson. In addition to the bequest to the minister, Fellgate's will bequeaths "one skype of beese" to his niece Ledea.

was, however, usually just of money.[22] A hive, of course, had a financial value; an inventory of goods from Writtle taken in 1638 assessed the widow Joane Harris's "2 heves of bees" at 6s 8d, for example.[23] But Fellgate's bequest, which linked bees with a sermon, gives us a hint that a hive had more than just an economic importance, and we can turn to beekeeping manuals to more fully understand this.

In his *Foure bookes of husbandry*, Conrad Heresbach wrote:

> This little poore creature the Bee, dooth not only with her labour yeelde vnto vs her delicate and most healthy Honey, but also with the good example of theyr painefull [i.e., careful] diligence and trauaile, encorageth man to labour and take paines, according to his calling: in such sort, as it seemeth the almightie and most excellent maiestie, hath of all other specially created this little poore creature, for the benefite and commodity of man: by whom, besides the commodity of the Hony and Waxe that they make, we might take both example to spend our life in vertuous and commendable exersises, and also to honour and reuerence the wonderfull bounty & goodnesse of the most gracious LORD shewed towards vs, in the creation of this small and profitable worme.[24]

Just as dairy cows were profitable for the production of milk and calves, so a hive produced honey and wax. But according to Heresbach, bees were also a moral example and pointed to God's generosity. Such an assessment allows us to speculate that a skep bequeathed to a vicar was not simply intended to

---

22. In 1629, two of Fellgate's neighbors, Richard Colle and Richard Whale, bequeathed, respectively, "Master Jefre the minister tenne shillinges vpon condicion he shall macke a sarmonn at my fenerall" and "my minister Mr Jeffery ten shillings desiring him to preach at my funeral." ERO, D/ACW 11/42 (1629), made probate 14 days after writing; and ERO, D/ACW 11/78 (1629), made probate 16 days after writing. The wills of Colle, Whale, and Fellgate were written by the same scribe, who might have been John Jefre the minister himself. Five other wills from Ramsey made probate between 1623 and 1632 are in the same hand, but none of them includes a bequest to Jefre. Other wills that include bequests for sermons include ERO, D/ACW 8/261 (1620), made probate 56 days after writing; ERO, D/ABW 47/221 (1625), made probate 515 days after writing; ERO, D/ACW 10/26 (1626), made probate 38 days after writing; and ERO, D/ACW 12/12 (1633), made probate 85 days after writing. Ilana Krausman Ben-Amos notes that bequeathing money for a funeral sermon "enhanced the credit and reputation of the deceased and celebrated their generosity"; see Ben-Amos, "Gifts and Favors: Informal Support in Early Modern England," *The Journal of Modern History* 72, no. 2 (2000): 22.

23. The inventory is in Steer, *Farm and Cottage Inventories*, p. 74. Harriss's will does not include bees as a specific bequest: they are included in the phrase "all other my goodes not bestowed" that she left to her son James, her executor. ERO, D/APwW 1/9 (1638), made probate 159 days after writing. None of the sixteen inventories associated with wills in the large ERO dataset include bees.

24. Heresbach, *Foure bookes of husbandry*, 174r.

supply his church with wax for candles.[25] Tracking the otherworldly reasons for this legacy uncovers a new way of thinking about bees and about the nature of early modern husbandry more generally.

Bees were distinguished from other quick cattle in wills in that they were the only creatures that were always represented collectively.[26] They were described variously in bequests in the large ERO dataset as "a skepe of beese,"[27] "one swarme & hyve of bees,"[28] "one heeve of been,"[29] "one hive of beese,"[30] "a stock of beese,"[31] "the baste heueses of bees in the gardene,"[32] and as "one great skep."[33] Their presence in wills as a colony rather than as individual insects makes sense—the alternative would be ridiculous and impractical, to say the least. Indeed, Edmund Southerne, writing in his 1593 beekeeping manual about tithes, argued that "I thinke Bees will hardly be profitable to the Parson, if he should haue but the tenth Bee: but the tenth part of the money which thou takest for sale of Bees, and the tenth part of

---

25. Walker and Crane note that "until the Reformation[,] the Church received many gifts of bees-wax, which was mostly used for candles and other lights" but that "after the Reformation, churches used much less beeswax." Walker and Crane, "English Beekeeping from c. 1200 to 1850," pp. 10–11.

26. Sheep were not named but were sometimes individualized ("my brittle face sheep being pold"); and even birds were sometimes counted and so given some basic individual status in the large ERO dataset. So although fourteen of twenty-two wills (64 percent) that included birds simply left "all my poultry" (or variations of that phrase), two testators (9 percent) were specific: the widow Elizabeth Sewell of Danbury bequeathed to her daughter Jane "my three hens & to Francis [another daughter] my three geese" and the widow Ann Birch left "one hen" to Joane Bracke, another widow. ERO, D/ABW 45/151 (1624), made probate 77 days after writing; ERO, D/ACW 11/222 (1631), made probate 11 days after writing. Other poultry bequests were more general. For example, the widow Julian Burre of Thorrington bequeathed "To Goodwife Mountford a cupple of henns and an other cupple of henns to Goodwife Painters wife [sic]"; and Humphrey Branwood of Woodham Mortimer's will included the following: "Item I give and bequeath vnto Elsabeth Branwood thother halfe of my Paulter my meaninge is that all my said Paulter shalbe equally devided between my son Edward & my daughter Elsabeth wch was mien." ERO, D/ABW 43/11 (1621), made probate 93 days after writing; and ERO, D/ABW 50/257 (1630), made probate 43 days after writing.

27. ERO, D/ACW 11/102 (1630), made probate 204 days after writing.

28. ERO, D/ABW 43/191 (1620), made probate 30 days after writing.

29. ERO, D/ABW 46/237 (1624), made probate 10 days after writing.

30. ERO, D/AEW 17/137 (1623), made probate 212 days after writing.

31. ERO, D/AMW 1/34 (1623), made probate 850 days after writing.

32. ERO, D/APbW 1/28 (1634), date of writing not recorded. This testator's burial is not recorded in parish registers for either Kelvedon or Kelvedon Hatch.

33. ERO, D/ABW 48/52 (1627), made probate 56 days after writing. This will—that of Hester Woulfe, a widow from Rickling—also included a reference to a basin "wch they vse to rynge bees with-all"; that is, a basin that was used to make a noise that was believed to encourage swarming. Woulfe included the "skep" in a list that included pothooks, trenchers, basins, and bottles and not in the later list of "all my fowles about the yard, excepting thre pullets) my greatest sowe & six piggs, my white cowe & my redd wennell, & twentie sheeps, some ewes & some hoggates." Given the conventional arrangement of the will—with a list of household things followed by a separate list of quick cattle—it is possible that the skep was empty.

the Honey and Wax which thou takest thy selfe, in conscience is due to the parson."[34]

The collective nature of bees in testamentary bequests reflects their representation in beekeeping manuals. Southerne, for example, noted that "when there are too many Drones in a Hiue . . . then the Bees of themselues will kill so many as they thinke good, so that I haue seene at least a pint lye dead vnder a hiue at once."[35] The measurement of the number of dead creatures ("a pint") represents them in terms of volume rather than as multiple individuals. And in Richard Remnant's *Discourse or Historie of Bees* of 1637 there appears to be a significant, if unnoticed, linguistic slippage from a focus on individual creatures to a focus on the hive that reinforces the sense of the animals' lack of individuality: "For choice of the best for store," Remnant writes, "alway keepe them that have the fairest and the evenest work, neither too fat nor too leane, and full of Bees."[36]

Alongside such collective representation, however, there was also understood to be a hierarchy in the hive that differentiated its members: the "master Bee" was at the top until 1609, when he was replaced by the "Queene" after Charles Butler proposed that a female was the ruler,[37] and Southerne wrote of the underlings that "although the Drones labour not abroade yet that which the other Bees bring home, they doe both helpe to work into Combes, and also to vnloade the Bees of their burthens, so that the work within is as necessarie as the others abroad."[38] Presenting each member as having a fixed place in the colony meant that there was little need for distinct characterization (the queen was the exception, but even she was usually treated as a bee that had a role rather than as an individual), and it also meant that the hive offered commentators what Mary Baine Campbell has called a

---

34. Edmund Southerne, *A Treatise Concerning the right vse and ordering of Bees* (London: Thomas Orwin, 1593), D4v. See the case that came before the King's Bench in 1635, which stated that a "tithe of bees may be discharged by a custom to pay tithe of their honey, wax, and the expences of hiving them." 79 E.R. 951, reported in Sir George Croke, *King's Bench Reports tempore Charles*, p. 951. Downloaded from westlaw.co.uk, accessed 24 May 2016. Lisa Jean Moore and Mary Kosut clarify the practicality of Southerne's understanding; they write that "honeybees do not survive as individuals. They must be part of a colony." Moore and Kosut, "Among the Colony: Ethnographic Fieldwork, Urban Bees and Intra-Specific Mindfulness," *Ethnography* 15, no. 4 (2014): 517.

35. Southerne, *A Treatise Concerning the right vse and ordering of Bees*, B4v.

36. Richard Remnant, *A Discourse or Historie of Bees* (London: Thomas Slater, 1637), p. 31.

37. Charles Butler, *The Feminine Monarchy or Treatise Concerning Bees* (Oxford: Joseph Barnes, 1609). Frederick R. Prete notes that Butler's work "came to be regarded as the first scientific book about honey bees and beekeeping." Prete, "Can Females Rule the Hive? The Controversy over Honey Bee Gender Roles in British Beekeeping Texts of the Sixteenth-Eighteenth Centuries," *Journal of the History of Biology* 24, no. 1 (1991): 127.

38. Southerne, *A Treatise Concerning the right vse and ordering of Bees*, B4v.

"naturalized concept or a 'natural' model" of political order.[39] The bees in beekeeping manuals were an example of an ideal commonwealth. This mirrored how they were represented in emblem books—that is, in poetic works that offered moral lessons to those who interpreted them.[40] In Geffrey Whitney's late-sixteenth-century collection *A Choice of Emblems*, for example, the orderliness of the social structure of the hive was celebrated as a model for human society in the emblem titled *"Patria cuique chara"* (home is where the heart is). There, the image of the hive was followed by a verse that included the lines "the Lorde in highe estate dothe staye: / By whose supporte, the meaner sorte doe liue."[41] The social structure in the human world was thus presented as natural and thus true and unchangeable.

Their presenting a useful model for human culture was evident not only in the manuals' discussions of the social structure of the hive, but also in the image of bees' *"busie industry"*[42] Butler described them as "enemies . . . to idlenes,"[43] and Thomas Hill wrote that "Such as be sluggish and not laboring [other bees] diligently note, which for their sluggishnes they biterly punish to death."[44] The fact that manual writers recognized that their industry was innate rather than needing to be trained into them by humans underlines the fact that the manuals' discussions of bees' busyness were not included as advice on how it might be produced. Instead, the manuals used the insects' industry to speak to other concerns. This, as Heresbach noted, was bees' value as exemplars. The manuals highlighted the emblematic rather than the practical relationship with them; they read them as natural models that point humans to correct (that is, to godly) behavior.

But, rather than reading the emblematic use of bees as utterly distinct from the lived relation, such moments in the manuals can and should be interpreted as also engaging with husbandry practices; as another means of addressing how to collaborate with these creatures that happened to use conventional moral terminology. The manuals, after all, were written

---

39. Mary Baine Campbell, "Busy Bees: Utopia, Dystopia and the Very Small," *Journal of Medieval and Early Modern Studies* 36, no. 3 (2006): 623. On the hive as a "generative metaphor for individual sacrifice for the benefit of the collective body of the kingdom," see Nicole A. Jacobs, "Bees: The Shakespearean Hive and the Virtues of Honey," in *The Shakespearean International Yearbook* 15, edited by Tom Bishop and Alexa Huang (Farnham: Ashgate, 2015), p. 102.

40. See Ashworth, "Natural History and the Emblematic Worldview," pp. 310–11.

41. Geffrey Whitney, *A Choice of Emblems* (1586), introduction by John Manning (Aldershot: Scolar Press, 1989), pp. 200–201.

42. Gervase Markham, *Cheape and Good Husbandry* (London: T. S., 1616), p. 172.

43. Butler, *Feminine Monarchy*, sect. 1.56, n.p.

44. Thomas Hill, *A Profitable Instruction of the Perfect Ordering of Bees* (London: Edward Alde, 1593), p. 5.

by people who worked closely with them: John Levett praised Southerne's work because, he argued, it contained "onely certaine practises and experiments gathered by his owne observations, by long keeping of Bees."[45] And the bringing together of worldly and otherworldly conceptions of the bees is evidenced in the expertise of the authors in another way too: it is not accidental, I think, that both Charles Butler and William Lawson (who referred to himself as "a Bee master") were clergymen.[46] So, a parson might possess a hive, and a beekeeping parishioner, who might not be able to read him or herself, could hear bees being used in a sermon in a way that linked their moral meaning with his or her practical understanding.

The link between the moral and the practical is not always easy to trace, however, and the existence of a relationship between husbandry and symbolism seems to be especially difficult to locate when bees' moral uprightness is praised. Thus, Lawson noted that they "of all other creatures, loue cleanlinesse and peace,"[47] and Butler wrote that "in the pleasures of their life the Bees are so moderate, that perfect temperance seemeth to rest only in them."[48] Such claims seem to present bees as natural models for humans and to lift us out of the world of husbandry, but this is not the only role of these statements. Butler, for example, took his sense of bees as the epitome of moderation into the realm of practical advice when he showed that it was in the interaction between them and their keepers that their virtues were made manifest:

> if thou wilt haue the favour of thy Bees that they sting thee not, thou must avoid such things as offend them: thou must not be (1) vnchast or (2) vncleanly: for impurity & sluttishnes (themselues being most chast and neate,) they vtterly abhorre: thou must not come amōg thē (3) smelling of sweat, or having a stinking breath caused either through eating of leekes, onions, garleeke, and the like; or by any other meanes . . . thou must not be givē to (4) surfeting & drunkennes.[49]

---

45. John Levett, *The Ordering of Bees: Or, The Trve History of Managing Them* (London: Thomas Harper, 1634), p. 3.

46. William Lawson, *A New Orchard and Garden* (London: Bar: Alsop, 1618), p. 19. Lawson is described as a "writer on gardening and Church of England clergyman" in John Considine, "Lawson, William (1553/4–1635)," in *Oxford Dictionary of National Biography* (Oxford: Oxford University Press, 2004). It is notable in this context that in March 1647, Ralph Josselin recorded bringing "a skippe of bees" to his vicarage in Earls Colne. Ralph Josselin, *The Diary of Ralph Josselin, 1616–1683*, edited by Alan Macfarlane (Oxford University Press, 1976), p. 90.

47. Lawson, *A New Orchard and Garden*, p. 21.

48. Butler, *Feminine Monarchy*, sect. 1.49, n.p.

49. Butler, *Feminine Monarchy*, sects. 1.49 and 1.33, n.p.

Estienne, similarly, told his reader that "before he handle the Hiues, he shall be well aduised, that the day before he haue not had to deale with his wife; that he haue not beene drunken."[50] And Markham noted that the bee "is a creature gentle, louing and familiar about the man which hath the ordering of them," adding pointedly, "so he come neate, sweet, and cleanly amongst them."[51] By logical implication, being stung by a bee was the mark of a moral failing on the part of the keeper.[52] Tending these creatures successfully was felt to require a personal hygiene that was understood as both moral and physical. The bees thus taught their keepers moderation—taught them, in short, to be good Christians. They were truly "living characters in the language of the Creator."

But it must also be recognized that this teaching by the bees was not just moral—a fact that is central to this chapter. Current beekeeping practice has shown that a keeper's change of shampoo can have an impact on a hive because of the bees' sense of smell,[53] and this was an understanding that existed 400 years ago. Remnant noted that bees had an "exceeding quicke" sense of smell, and were able to trace scents from over a mile away.[54] Such recognition reveals that early modern assessments of bees, however moralizing their language, were not simply emblematic. It shows that the authors of beekeeping manuals of that period recognized real encounters and understood that a person's odor could actually affect the animals' behavior. Thus, the descriptions of bees' response to their postcoital or hungover keeper should be read as revealing early modern writers' recognition that what underpinned the encounters between people and their hives was the existence of the animals as responding partners. Indeed, the recommendations about cleanliness that were repeated in so many beekeeping manuals show that Vinciane Despret's argument that animals have opinions is not wholly new.[55] Early modern writers acknowledged that keeping bees required the keepers to attend to the bees' behavior and respond to it appropriately, just as milkmaids needed to consider

50. Estienne, *Maison Rustique*, p. 322.

51. Markham, *Cheape and Good Husbandry*, p. 172.

52. This is reminiscent of the early modern understanding of illness, which regarded it as a marker of the moral failure of the sufferer. See Michael C. Schoenfeldt, *Bodies and Selves in Early Modern England: Physiology and Inwardness in Spenser, Shakespeare, Herbert, and Milton* (Cambridge: Cambridge University Press, 1999), p. 7.

53. Personal communication from Rebecca Marsland. On bees' sense of smell, see Eileen Crist, "Can an Insect Speak? The Case of the Honeybee Dance Language," *Social Studies of Science* 34, no. 1 (2004): 10.

54. Remnant, *A Discourse or Historie of Bees*, p. 11.

55. Vinciane Despret, "Sheep Do Have Opinions," translated by Liz Carey-Libbrecht, in *Making Things Public: Atmosphere of Democracy*, edited by Bruno Latour and Peter Weibel (Cambridge, MA: MIT Press, 2005).

their apparel when milking because they understood the cows' potential to respond to that. And beekeepers were felt to need advice because beekeeping was believed to be difficult because it had to deal with the bees and their point of view.[56] In this context, knowing that bees lived within a fixed hierarchy, that they were led by one individual, that they responded to new smells, was not just of possible emblematic value; it was also important in practical terms if the beekeeper was to be successful. The fact that particular social or moral meaning was attached to such aspects of beekeeping in early modern manuals does not detract from their containing recognition of actual behavior by actual bees.

The co-presence of emblematic and practical meaning in beekeeping manuals, then, reveals a number of things. It shows that the emblematic worldview was not just made in libraries but was underpinned by empirical understanding. This is not saying anything that has not been noted before.[57] Most important here, though, it also illustrates that there was a relationship between the material and immaterial realms when it came to working with animals. Bees, after all, were recognized in early seventeenth-century husbandry manuals as being simultaneously producers of honey, wax, and meaning. And the simultaneity is crucial. The people who owned and worked with bees regarded husbandry as a process with more than one engaged party and were making sense of the animals they worked with in ways that took their animals' behavior seriously. They were, to invoke the ideas of historians of science Carla Hustak and Natasha Myers, "involving" themselves in the lives of their bees and allowing the animals to involve themselves in their (human) lives.[58] Their keepers thus knew that bees would respond to certain odors and allowed that knowledge to impact upon their own (human) behavior. Thus, calls for temperance in the manuals did not simply read the bees as emblematically valuable—although of course they were doing that; they also offered solid agricultural advice: don't smell bad, respect that your bees have

---

56. Such an assessment was also available in classical ideas. Plutarch noted that the "in bred sagacity" of "sea creatures" meant that "the arte of hunting and catching them is not a small piece of worke, and a simple cunning; but that which requireth a great number of engins of all sorts, and asketh woonderfull devices." Plutarch, "Whether Creatvres Be More Wise, They of the Land, or Those of the Water," in *The Philosophie, commonlie called, The Morals* (London: Arnold Hatfield, 1603), p. 970.

57. See Brigitte Resl, "Introduction: Animals in Culture, ca. 1000-ca. 1400," in *A Cultural History of Animals in the Medieval Age*, edited by Brigitte Resl (Oxford: Berg, 2007), pp. 10–15.

58. Hustak and Myers challenge what they term "the reductive, mechanistic, and adaptationist logics that ground the ecological sciences." They celebrate instead "accounts of the creative, improvisational, and fleeting practices through which [different partners] *involve* themselves in one another's lives." Carla Hustak and Natasha Myers, "Involutionary Momentum: Affective Ecologies and the Sciences of Plant/Insect Encounters," quoted in Vinciane Despret, "From Secret Agents to Interagency," *History and Theory* 52, no. 4 (2013): 34.

the capacity to respond and that their responses must be factored into your husbandry and your smallholding will be more successful.

When read in this context, the emblematic worldview can be seen as constituted by as well as constituting practical experience. So it should come as no surprise that when Ralph Josselin, the beekeeping vicar of Earls Colne, was stung in September 1644, he found both a lesson and a cure in the hive:

> Stung I was with a bee on my nose, I presently pluckt out the sting, and layd on honey, so that my face swelled not, thus divine providence reaches to the lowest things. lett not sin oh Lord that dreadfull sting be able to poyson mee.[59]

What emerges from the intertwining of the symbolic and the practical is that bees must be recognized as having both worldly and otherworldly value; must be understood to offer their keepers useful stuff but also opportunities for self-reflection. And this might have been what John Fellgate had in mind when he left his minister the bequest of a hive in exchange for a funeral sermon. Even bees, those tiny, non-individualized creatures, had meaning beyond their financial value.

But bees are not the only creatures that offer a glimpse of the presence of the emblematic worldview in the large ERO dataset. Bequests of lambs also reveal a link to transcendent matters. In addition, they take us toward the problem that the slaughter of animals with whom people had close relationships created. To get to this crucial issue, however, we need to begin by contemplating more generally who got what in a will.

As the wills of Robert Jacobb and Christopher Fuller showed, members of the family were the main legatees in wills, although household staff— Jacobb's "man" and perhaps Alce Milles—also received gifts. But other wills reveal a wider community of beneficiaries that stretches beyond the immediate household to include neighbors, siblings, cousins, and those named as kin more generally. Clearly, testators made choices about who to recognize and how to recognize them, and these choices are revealing. In the sample dataset of eighty-nine wills, for example, seventy-one (80 percent) include bequests of any kind—property, money, goods, or animals—to the sons and daughters of the testator. It makes sense, of course, that the percentage of wills including bequests to the testator's own children would be so high. For a widowed parent, as Fuller's will showed, sons and daughters were the obvious beneficiaries of a lifetime's accumulated wealth. Even in the fifty wills in

---

59. Josselin, *The Diary of Ralph Josselin*, p. 19.

the sample in which the surviving spouse received a bequest or was named as the executrix, thirty-nine (78 percent) also included legacies for their children.[60] In support of this, it is notable that childless testators were more likely to distribute their goods widely than those with offspring: of the thirty-four wills in the sample dataset that include bequests to individuals who were not members of the testator's nuclear family, only eighteen (53 percent) also included gifts the testator's own children.[61]

Earlier chapters have argued that livestock made up a significant part of many individuals' wealth, and thus it makes sense that quick cattle, like property, money and non-self-moving movable goods, would be left to offspring. This is visible in the 451 wills that include specific animal bequests; 212 (47 percent) left quick cattle to the testator's children.[62] The percentage is lower than the percentage of wills that bequeathed property in general to children (80 percent of the sample), perhaps because looking after animals was more likely to be undertaken by adults. However, as in the sample dataset, wills that do not include legacies of animals to sons or daughters and are plausibly from a childless testator are much more likely to distribute quick cattle widely, thus reinforcing the sense of the priority given to immediate offspring where they exist. Of the sixty-five wills that bequeath livestock to those I am terming neighbors (individuals who are not named as relatives and do not share a family name with the testator) just nine (14 percent) also include bequests of animals to the testator's sons or daughters. The figure is even smaller for wills that include bequests of beasts to a wider kin network (brothers, sisters, uncles, cousins, and individuals simply termed "kinsman" or "kinswoman"): only five (8 percent) of the sixty-four wills that leave animals to such individuals also include animal bequests to the testator's own children. And in none of the ten wills that include bequests of animals to one or both parents did the testator leave a beast to a child.[63] By implication, where children were present, they were more likely to receive

---

60. Some wills include a direction that the spouse—usually a wife—should keep possessions for their young children.

61. On the distribution of property in wills, see David Cressy, "Kinship and Kin Interaction in Early Modern England," *Past and Present* 113 (November 1986): 38–69; Ralph Houlbrooke, *Death, Religion, and the Family in England, 1480–1750* (Oxford: Oxford University Press, 1998), pp. 140–42. Richard T. Vann's data offers alternative readings of the spread of kin recognized in wills; see Vann, "Wills and the Family in an English Town: Banbury, 1550–1800," *Journal of Family History* 4, no. 4 (1979): 363–66.

62. I am including bequests to individuals named as daughters-in-law and sons-in-law and those named as a wife's son or daughter in these figures. One hundred eighteen wills (26 percent) leave specific animal bequests to wives. This figure does not include those that make the wife the will's executor and by default leave her all animals not named in the will.

63. Such wills may be by young testators, and while none specifically bequeaths an animal to a wife, two mention their wife as the will's executrix.

livestock than anyone else. But because care for the animals was also a possible consideration in bequeathing patterns, testators may have considered who could use and look after animals right away. Most commonly, this was the testator's spouse.

Given the important role that quick cattle might play in a testator's planning for the future after his or her death, it seems to be saying little, therefore, to emphasize the fact that children received animal legacies from their parents. But it was not just the testator's own children who received them: members of the generation younger than the testator more generally were common recipients of quick cattle at an individual's death. If we widen the category "children" to include those named as nephews,[64] nieces, grandchildren, and godchildren and legatees specified as the children of other neighbors or kin (in what follows, where the word children appears inside quotation marks, it signifies this wider grouping), then the proportion of wills that left any bequest to such "children" in the sample dataset is 89 percent (79 of 89 wills). In the set of wills that include specific animal bequests it is 64 percent (288 of 451 wills). The pattern can also be found when we turn to the bequeathing of what are named as specifically young animals, and it is here, I suggest, that we might begin to uncover another emblematic potential of animal bequests.

Of the ninety wills in the large ERO dataset that include bequests of lambs, sixty-five (72 percent) have legatees who are "children," and sixty-nine of the ninety-eight wills (70 percent) that include young bovine animals (heifers, steers, bullocks, weanels, calves[65]) leave them to legatees in this group. Of the forty-four wills that include colts, twenty-one (48 percent) leave them to

---

64. Naomi Tadmor notes that "nephew" in this period could be used to designate a grandchild. However, when the term "nephew" is used in Essex wills it appears to describe a sibling's child. For example, Samuel Cooper's married sister was named Mary Hayward and his will referred to Steven Hayward as "his nephew" see ERO, D/ACW 11/142 (1632), made probate 62 days after speaking. Tadmor, "Early Modern Kinship in the Long Run: Reflections on Continuity and Change," *Continuity and Change* 25, no. 1 (2010): 32.

65. The grouping is obviously problematic, given the confused usage of these terms: animals described as heifers had calves and bullocks were much older than might be expected, and so on. For want of alternatives, I have taken these terms to refer to young (or younger) bovine animals than those described as bulls or cows. "Steer" is used only in the will of Christopher Fuller. The terminology used to describe swine is also less than clear; but I have taken the terms pig, piglet, and shots here to signal young animals. The meaning of "lamb" seems much simpler: many were bequeathed alongside their mother, thus hinting at their immaturity. Other descriptions differentiate a "ram lamb" from a "ewe lamb" or refer to age and color. For example, Richard Walker of Buttsbury bequeathed "one Lambe of a yeere Oulde"; Thomas Whitnall, a yeoman of Goldhanger, left "a black year ould Lam"; and Silvester Pett, a farmer from Mundon (one of only four "farmers" in the dataset) left "one Russett Ewe lambe." See, respectively, ERO, D/ABW 44/11 (1622), made probate 3,257 days after writing; ERO, D/ABW 48/113 (1627), made probate 46 days after writing; ERO, D/AEW 19/299 (1634), made probate 50 days after writing.

"children,"[66] and of the twenty wills that refer to pigs (piglets) or shots, nine (45 percent) do the same.[67] These figures suggest a number of things. It is plausible that except at the highest end of the social scale, horses would be more likely to have adult riders and users and that testamentary practices reflect this. Pigs, as we have seen, were not often specifically referred to in wills, but here, when they are, the lower than average proportion of "child" legatees might show that young pigs may have been regarded as somewhat independent residents of the yard: feeding from their mother's milk, then foraging for much of their own food, and otherwise living off scraps. Caring for these creatures may have been understood to be different from looking after other young animals—to have required less exercise of skill—and this could have had an impact on who was bequeathed them. The fact that of all animals, pigs are most likely to have been slaughtered at home might also play a role here. I return to this.

We can drill down even further into the bequeathing of young animals to young legatees in the Essex data to gain a fuller comprehension of the practice by looking at how testators described the individual legatees (the previous statistics were based on individual wills rather than on particular legatees). Here a clear difference in bequeathing patterns can be traced. All fourteen of the legatees who fall into the category "children" who were left young swine were the immediate offspring of the testator. Of the twenty-one "children" who were bequeathed colts, twenty (95 percent) were sons and daughters of the testator, and of the ninety-three individual "child" legatees who were left young bovids, seventy-three (78 percent) were named as immediate offspring of the testator.[68] However, only thirty-five of 123 "children" (28 percent) who were bequeathed lambs were the sons

---

66. Gervase Markham claimed that colts were separated from the mare at one year, whereas Peter Edwards notes that other manuals stated that foals should not be weaned until they were three years old. Markham, *Cauelarice: Or the English horseman* (London: Edward Allde, 1607), pp. 63–66; Edwards, *Horse and Man in Early Modern England* (London: Hambledon Continuum, 2007), p. 39. The variation in use of the term "colt" is evident in the large ERO dataset, which includes bequests of one four-year-old colt, one three-year-old colt, one two-year-old colt, and one eighteen-month-old colt. The will of the ranger John Mylborne of Havering Park mentioned three colts: "my baye Coulte of fower yeare olde . . . one white graye Coulte of a yeare olde [and] A Dune Mare and her Coulte"; ERO, D/AEW 18/43 (1626), made probate 639 days after writing. The loose use of the term is evident in the will of Thomas Pole of Little Clacton, who mentioned "our Mare Colt with a bald fface, now bigg with Colt"; ERO, D/ABW 47/15 (1626), made probate 113 days after writing. The term "foal" is not used in any will.

67. I have not included poultry in this survey because so many wills (13 of 22) use only the generic terms "poultry" or "fowls." Only one of the four wills that specify young poultry left them to a child (25 percent).

68. This figure includes four individuals described as the "wife's daughter," two sons-in-law, and one daughter-in-law. At this point, these individuals—some of whom are obviously adult—are

and daughters of the testator.[69] Lambs were thus much more likely than all other young animals to be left to children who were not the testator's immediate offspring.

The distinctive nature of the pattern of bequeathing lambs can be tracked even further by looking at the age of "child" legatees. In a will, a "child" listed as married or as having their own children was clearly an adult, for example, and parish registers can be used to confirm the age of a legatee, assuming the date of baptism to be close enough to the date of birth to act as an equivalent. From these sources, it is possible to establish the age of a proportion of the legatees who fall into the wide category "children" and thus to track with even more specificity how young animals were bequeathed. So, of the ten "children" who were bequeathed young pigs, the age of four cannot be confirmed, while the other six were all over the age of eighteen. In the group of twenty-one "child" recipients of colts, the age of eight cannot be confirmed, but of the remaining thirteen, twelve (92 percent)—including a grandchild—are confirmed as adults, with Nicholas Convin's fourteen-year-old son Edward the only recipient under the age of eighteen.[70] And, while the age of forty-three of the ninety-three "children" who were bequeathed young bovids cannot be determined, thirty-three of the remaining fifty (66 percent) were definitely over eighteen and seventeen (34 percent) were definitely under eighteen at the time the wills were written. The statistics in wills that include bequests of lambs are rather different once again. The ages of sixty-six of the 123 "children" who were left these animals cannot be confirmed, but of the remaining fifty-seven, thirteen (23 percent) were definitely adults at the time the will was written, while forty-four (77 percent) were under eighteen. In addition, the age of nineteen individual child recipients of young bovids, and twenty-six recipients of lambs can be confirmed by baptismal records.[71] The average age of a child who received a young bovid was approximately

---

included in my figures. It is likely that a daughter- or son-in-law was a legatee after the death of the testator's biological offspring, and so was the parent to the testator's grandchildren.

69. The only "child" who was bequeathed pullets (young chickens) was a granddaughter. Her mother received the mature birds in the same will. See ERO, D/ABW 48/52 (1627), made probate 56 days after writing. Because birds mature so rapidly, it is likely that the distinction between pullets and hens was less meaningful than between, say, sheep and lambs.

70. ERO, D/ABW 48/152 (1626), made probate 86 days after writing. Edward lived for less than two months after his father Nicholas had written his will so had little time to enjoy his legacy. Parish Register, St Mary the Virgin, Mistley, ERO, D/P 343/1/1 (1559–1696).

71. The reason the number of children whose exact age can be confirmed is lower than those who were under eighteen in wills that bequeathed young bovines and lambs to children is that I have included in the latter groups siblings of individuals whose age can be confirmed or have used the date of the marriage of their parents to confirm the earliest date of birth of individual legatees who are their offspring in the absence of the date those legatees were baptized.

eight years and eight months; the average age of a child who received a lamb was approximately five years and five months. The sample sizes are small, but the difference, again, is notable.

On the basis of these statistics, it is worth asking why lambs were so much more likely than any other young animal to be bequeathed to children who were under the age of eighteen and why it appears that those children might have been younger on average than the children under eighteen who received young bovids. Issues of practicality are obviously crucial. While it was possible to leave a young child a full-grown cow, of course, it is unlikely that the child would have been expected to take care of that animal him or herself. Indeed, often the cow was noted as not to be delivered to the young legatee until they reached adulthood. For example, Thomas Burre of St Osyth bequeathed "to Thomas Burre John Burre and Richard Burre to evrie his sonnes to eurie of them one Cowe to be delivered to them at their seurall ages of one & twenty yeares."[72] What was being left in such a bequest was not so much the animal's value to the legatee at the moment of writing the will as its potential for future usefulness. This was also the case when young bovids were bequeathed, something that was acknowledged by John Wallys, a husbandman from Great Chishill, who bequeathed to his son Thomas "mye red bullock (beyinge a two yere ould bullock at the next springe) hopinge that mye wiff aforesayde will keepe and bringe vpp the same bullock vntill some profitable Increase therebye cometh to mye sonn Thomas."[73] While a bullock such as this might be destined in time for the market as meat and might thus represent a short-term gain with the sale price given to the child, the young animal might also have the potential to become part of a dairy herd (creatures termed "bullocks," as we have seen, could be female). This would mean that it was possible that this bullock would have a role to play in the child's upbringing in terms of both the nutritional value to the household of her milk and the economic value of her calves. It is this that might be her "profitable Increase." In addition, if a dairy cow could work for up to fifteen years (as Leonard Mascall estimated), what was being bequeathed could also be a longer-term investment in the child's future. Thomas Wallys was eleven years old when his father died, and

---

72. ERO, D/ABW 51/244 (1632), made probate within 74 days of speaking. No register exists that would allow us to trace any of the sons' exact ages. William Biatt of Stebbing bequeathed his son John "a Cowe to be delivered vnto him I meane the Cowe when he goeth to keep house"; ERO, D/ABW 48/41 (1627), made probate 55 days after writing. No register exists that would make it possible to trace John's exact age.

73. ERO, D/ACW 8/296 (1621), made probate 75 days after writing.

it might be that he would use the not yet two-year-old bullock as the start of his own dairy herd in years to come.[74]

The potential for long-term productivity also holds true for sheep. As previously noted, Thomas Tusser advised that eating lambs was poor financial planning, that the profit in them was in wool, not meat.[75] Likewise, Peter J. Bowden cites the mid-seventeenth-century farmer Henry Best's note that he "disposed of his ewes before they reached the age of eight but many men . . . found it profitable to keep them for several years longer."[76] For this reason, young male lambs not required for reproductive purposes were castrated (such were termed "wethers") and were raised for wool and only finally for meat.[77] Ewe lambs had a future quadruple productivity: their wool, their production of lambs, their milk, and—ultimately—their flesh.[78]

In practical terms, then, those testators who bequeathed a lamb to a child were offering a legacy that would have been of small value at the time of writing the will but had potential for profit over the long term. Thus, when Robert Gobby a clothier of Langham left his namesake grandson "a lambe, or fiue shillings in monye," this might not have been the value of the animal at the moment of bequeathing.[79] Indeed, an inventory attached to a will from a year before Gobby's valued "3 sheepe" at 15s, implying that the financial alternative that Gobby offered might represent the value of the adult animal rather than the young one.[80] With lambs, as with some young bovids, testators were bequeathing creatures who might provide that child with a potential income

---

74. Thomas's baptism is recorded in the Parish Register of St Swithin, Great Chishall, ERO, D/P 210/1/1 (1583–1747).

75. Thomas Tusser, *Five Hundred Points of Good Husbandry* (London: I. O., 1638), p. 60.

76. Peter J. Bowden, *The Wool Trade of Tudor and Stuart England* (1962; repr., London: Frank Cass, 1971), p. 21.

77. It is notable that three of four bequests of rams were to members of the same generation as the testator (two rams to the wife of a testator and one to a male cousin) and three of five bequests of ram lambs were also to this group (one to a male neighbor, one to a brother, and one to a brother-in-law of the testator). An individual from the younger generation received the vast majority of bequests of all other ovine animals—sheep, ewes, wethers, lambs, and ewe lambs.

78. William Camden noted that on Canvey Island there were "about foure hundered sheepe, whose flesh is of a most sweet and delicate taste, which I haue seene young lads, taking womens function, with stooles fastened into their buttockes to milke, yea and to make cheeses of Ewes milke in those dairy sheddes of theirs that they call there, Wiches." The same was true, he stated, of the sheep in Dengie Hundred. Camden, *Britain, or A chorographicall description of the most flourishing kingdomes, England, Scotland and Ireland, translated by Philemon Holland* (London: Iohn Norton, 1610), pp. 441 and 443. I have assumed that these were notable occurrences because men did the milking and that sheep's milk was commonly used for human consumption.

79. ERO, D/ABW 49/201 (1629), made probate 123 days after writing.

80. ERO, D/AMW 2/125 (1628), made probate 45 after writing. The value of a lamb might be visible in the 1626 will of Nicholas Peacock, a yeoman of Ulting, who left his unnamed godson 10 shillings to buy "three Lames to be brought vp vpon the common." ERO, D/ACW 10/26 (1626).

for a number of years, so the choice Gobby's grandson had to make was whether to take the short-term gain of a cash bequest or have faith in the survival and future extended productivity of the animal.

This sense of the long-term value of the lamb can be read in another way, too. When William Hutlie, a yeoman of Little Easton, laid down in his will that his wife Margaret "shall keepe for o'r saide sonne Euerie yeare during his nonage one sheepe and one Lambe" what he was leaving nine-year-old John was the beginning of his own flock.[81] But this bequest should not just be read as providing him with a valuable resource, it should also be interpreted as allowing John to learn his future trade on a small scale. Watching over the lambs—even if from a distance, as they were tended most closely by their mothers and, if needed, by a shepherd—would introduce this young legatee to the occupation he would be likely to enter into as he grew up, and might reveal the legacy to have had a hands-on rather than just economic value.[82]

But there was more to bequeathing a lamb than even these practicalities, and there is one place where I suggest a key emblematic meaning of the bequest becomes visible in the wills. Nineteen "child" legatees receiving lambs (over 16 percent) were named as godchildren of the testator.[83] By comparison, only two "child" recipients of young bovids were godchildren (2 percent of all such bequests to "children");[84] and no other young animal was bequeathed to a godchild. Thus, the bequeathing of lambs is yet again anomalous. This is worth pausing over as it is here, I suggest, that lambs emerge most clearly as, like pelicans and bees, "living characters in the language of the Creator."

---

81. ERO, D/ABW 47/185 (1625), made probate 977 days after writing. John's baptism was recorded on 7 February 1614 in the register of St Mary the Virgin, Little Easton, ERO, D/P 180/1/1 (1559–1783).

82. Bernard Capp records a case from Leicestershire in the 1630s that illustrates the danger of leaving a child in charge of animals. John Salter, who was employed as a shepherd, "was offered some better-paid harvest work and promptly abandoned the sheep to the care of a small child. When some animals were lost, the disgruntled owner charged Salter with stealing them or allowing them to be stolen, and had him sent to the Leicester bridewell and whipped." Capp, "Life, Love and Litigation: Sileby in the 1630s," *Past and Present* 182 (February 2004): 69.

83. Of those nineteen, twelve were male and seven were female. While the age of fifteen cannot be determined, the four whose ages can be traced were all under eighteen at the time the will was written.

84. The age cannot be determined for these two male godchildren. We have already encountered a godmother's gift of calves in the will of John Jeppes, who bequeathed to "Elizabeth Jeppes my daughter two bullocks the one black howed and the other a browne motly ~~which when they were calves hir godmother the wyddow Webb did give her~~"; ERO D/ACW 9/91 (1622), made probate 76 days after writing. Monetary bequests were made to godchildren in a number of wills. See ERO, D/ACW 8/266 (1620), made probate 53 days after writing; ERO, D/ACW 9/229 (1625), made probate 36 days after writing; and ERO, D/ACW 9/229 (1628), made probate 237 days after writing.

In Protestant England, the role of godparents was set out in the service for baptism in *The Boke of common prayer*, where they were asked to stand in the place of the newborn and declare the infant's faith on his or her behalf (in its adjectival use *infans* is Latin for "without language"). Godparents were also presented as the child's "suerties"—guarantors that they would learn the Creed, the Lord's Prayer, and Ten Commandments, and attend sermons, and that "that these childrē may be vertuously brought vp, to leade a godlye and Christen lyfe."[85] Godparents, in Richard Hooker's terminology, were "Fathers and Mothers in God."[86] In the theology of the Church of England, baptism did not bring about the child's possession of grace (no human act could do this), it symbolized it, and I suggest that the bequeathing of lambs to godchildren might have had a similar purpose. Thus when John Greene of Wormingford bequeathed to his "two godchildren Mary Greene (daughter of my sonne Willm) & George Pilgrym eyther of them a lambe,"[87] and when Robert Scarlet, a husbandman from Goldhanger, gave "unto my godsonne, the sonne of John Grant, one ewe-lambe,"[88] what they were giving was that surety in animal form.

This works on a number of intertwined levels—an intertwining that reveals, once again, the link between husbandry and symbolism in this period. First, and obviously, there is an unmistakable connection being made between the lamb and the child—not on the basis of innocence, however, but on the basis of practicality, with the lamb held up as an ideal infant. Topsell, for example, followed the tradition of twelfth-century bestiaries when he noted the lamb's ability to discern "the voice of his parent" when lost amid a flock of thousands.[89] This behavior, of course, can be observed in actual lambs.[90] But Topsell's Protestant readers were perhaps also being asked to recall the true (that is, natural, innate) obedience of that other lamb, Christ, and to bring to mind their own very human capacity to stray. Indeed, the absence of

---

85. "'The Ministracion of Baptisme,"' in *The Boke of common prayer and administration of the Sacramentes and other rites and Ceremonies in the Church of Englande* (London: Edwardi Whytchurche, 1552), Pviir and Pviiv. In addition, in England and Ireland the child might be given the name of the chief godparent; see Clodagh Tait, "Namesakes and Nicknames: Naming Practices in Early Modern Ireland, 1540–1700," *Continuity and Change* 21, no. 2 (2006): 320.

86. Hooker, *Of the Laws of Ecclesiastical Polity*, quoted in William Coster, "'From Fire and Water': The Responsibilities of Godparents in Early Modern England," *Studies in Church History* 31 (1994): 301.

87. ERO, D/ABW 47/141 (1626), made probate 25 days after writing.

88. ERO, D/ABW 48/218 (1626), made probate 143 days after writing.

89. Edward Topsell, *Historie of Foure-Footed Beastes* (London: William Jaggard, 1607), p. 640.

90. According to Brigitte Resl, twelfth-century bestiaries introduced a greater focus on domestic (rather than exotic) animals to the genre that had originated in the *Physiologus* of the third century CE. It was the twelfth-century texts that influenced natural philosophers in the seventeenth century. Resl, "Introduction," pp. 12–14.

links between young ovid and young human based on their shared innocence was likely to be meaningful to Topsell's readers because Protestant theology assumed the innate corruption of all humans from birth, a corruption that the individual Christian had no capacity to change: as Arthur Dent's Theologus noted, "We are by nature the children of wrath."[91] Thus, instead of appealing to their potential for self-perfectibility, Protestant authors urged their readers to put their faith in the forgiveness of the Creator, a faith that found its way into wills, as the Stansted Mountfitchet yeoman, John Eliott's makes clear: "I bequeath my soule into the m'cifull hands & p'tection of almightie God my creator and of Jesus Christ savior and redeemer of all mankinde, trusting assuredly by & throughe the meritts love & int'cession of the same Jesus Christ to have remission of all my sinnes & therby to obtayne eternall life."[92]

George Herbert, as we have seen, noted that a good country parson "condescends even to the knowledge of . . . pastorage . . . because people by what they understand, are best led to what they understand not," and at the conclusion of Topsell's "discourse of the Lambe," he notes that "the greatest honour thereof is for that it pleased God to call his blessed Son our Sauiour by the name of a Lamb in the old Testament, a Lambe for sacrifice; & in the new Testament, styled by Iohn Baptist, the Lambe of God that taketh away the sinnes of the world."[93] Such a familiar understanding might mean that the bequest of a lamb, in addition to having worldly significance, was giving the child an emblem of Christ. If that was the case, it was also a preparation for the death that existed in the midst of life and a marker of the child's potential to be saved. A godparent leaving a lamb to a godchild was thus an offering, as well as a valuable thing and the potential for training the child in their future profession, a kind of homely introduction to soteriology, the doctrine of salvation.

Alongside its practical value, therefore, the symbolic link of a lamb with Christ may have meant that this young creature rather than any other was felt to be the most appropriate to bequeath to a godchild. Young bovids, for example, did not have these relevant emblematic meanings; indeed, Topsell noted that calves are "exceedingly giuen to sport and wantonnes."[94] The bequest of a lamb, you might say, reinforced the relationship between godparent and godchild even as death was about to end it. In addition, it was offering the child a preparation for mortality in another way, and it is in this latter

---

91. Dent, *The plaine man's path-way*, p. 4.
92. ERO, D/ABW 44/267 (1623), made probate 70 days after writing.
93. Topsell, *Historie of Foure-Footed Beastes*, p. 641.
94. Topsell, *Historie of Foure-Footed Beastes*, p. 89.

context that we might see perhaps the most significant work of the emblematic worldview in the realm of husbandry.

Killing animals casts its shadow over all human-livestock relations; it is, as Jeremy McInerney has noted, the "paradox" at the heart of pastoralism. In distinction from hunter-gatherer societies, pastoral cultures, he writes, face a problem: farmers engage in what he represents as a parental relationship with their livestock in their overseeing of their breeding and rearing, and thus "killing a domesticated animal is tinged with betrayal in ways that would never trouble a hunter."[95] For the ancient Greeks, McInerney argues, sacrifice "resolve[d] the paradox of pastoralism": it gave meaning to the death of domestic animals by giving those deaths a sacred status.[96] In early seventeenth-century Essex, Protestant theology was, for most, the religion that was used to explain human mortality,[97] and writing a will played a role in the process of the individual's preparation for and acceptance of death, as we have seen. How or whether an early modern animal prepared for its death is not a question that can be answered, of course, but how the human who lived with, worked with, and cared for that animal prepared for its death is something we might be able to get a glimpse of. Indeed, the shadow cast by the necessity of killing the animals with whom the husbandman and his wife worked day-in, day-out is, I suspect, central to how they lived with them, and without understanding this aspect of husbandry, we cannot fully comprehend what husbandry was. With many animals, after all, keeping them alive and healthy was necessary only as a precursor to killing them.

Some quick cattle would be sold to butchers for slaughter to bring in an income, as we have seen, and butchers normally killed animals that were sold for meat in the marketplace. This was because they needed to provide evidence of the animals' health: in Malden, for example, local regulations stated that "if a butcher did 'kill his beast at his house, [it was necessary] that he bring his flesh to the market with the skin, that the Market Looker may have the oversight thereof, whether the flesh be able to be sold or not.'"[98] But the shift from home to market not only made the slaughter visible and thus

95. Jeremy McInerney, *The Cattle of the Sun: Cows and Culture in the World of the Ancient Greeks* (Princeton and Oxford: Princeton University Press, 2010), p. 36.

96. McInerney, *The Cattle of the Sun*, p. 40. Other kinds of meaning are given to the killing of hunted animals; see Tim Ingold, *The Perception of the Environment: Essays on Livelihood, Dwelling and Skill* (2000; repr., London and New York: Routledge, 2011), pp. 65–69.

97. On the religion of the county, see John Walter, "Confessional Politics in Pre-Civil War Essex: Prayer Books, Profanations, and Petitions," *The Historical Journal* 44, no. 3 (2001): 677–701.

98. Quoted in W. J. Petchey, *A Prospect of Maldon, 1500–1689* (Chelmsford: Essex Record Office Publications, 1991), p. 118. See also the tenth-century rule that "two witnesses are required when stock

turned the animal's death into an exhibition of well-being, it also distanced the people who were most closely engaged with the animal during its life from the event of its death, a distance that might have been important.[99] But where its flesh was kept for the household's own consumption (as the meat of many pigs on smallholdings would be), an animal had to be raised and then slaughtered within the domestic sphere, and the human need for nutrition would have to override any attachment to it. In "Decembers Abstract," Thomas Tusser simply noted, "Let bore life render,"[100] and explaining this annual slaughter of swine as just a part of the agricultural year is surely both right and wrong.[101] Killing a pig had to happen; the meat was salted and consumed over the winter months when other food was more scarce. But killing a pig is not the same as planting "willow and sallow" (another task Tusser listed for the same month) and is a job that might have sat uncomfortably alongside the care that was to be given at that time to cows (and the prospective care to be given to sheep):

> The housing of cattell, while winter doth hold,
> is good for all such as are feeble and old:
> It saveth much compasse, and many a sleepe,
> and spareth the pasture, for walk of thy sheepe.[102]

While clearly important, pragmatism—a belief that the slaughter of pigs must take place—does not answer the problem of how raising animals and killing them oneself might happen. This is not simply a question of the literal matter of who did the killing, although that was certainly important, with wealthier families, perhaps, always more distanced from acts of slaughter than their poorer neighbors could be.[103] And how the killing was carried out

---

is killed," quoted in Derrick Rixson, *The History of Meat Trading* (Nottingham: Nottingham University Press, 2000), p. 86.

99. This is still the case on contemporary farms. See Rhoda Wilkie, *Livestock/Deadstock: Working with Farm Animals from Birth to Slaughter* (Philadelphia: Temple University Press, 2010), pp. 129–46. I return to this issue in the afterword.

100. Tusser, *Five Hundred Points of Good Husbandry*, p. 48.

101. Derrick Rixson notes that Martinmas, 11 November, was the traditional date when the "autumn kill began." Rixon, *History of Meat Trading*, p. 96.

102. Tusser, *Five Hundred Points of Good Husbandry*, pp. 48 and 49. Sallow is a gray willow tree.

103. In 1631, for example, the "Household expenditure accounts" of the Petre family of Ingatestone Hall record four separate payments each of 4d "for one to kell a calfe." "Household expenditure accounts, probably kept by a member of branch of the Petre family [of Ingatestone]," ERO, D/DP A74 (1631–1640).

is also not the key issue at stake here, although, again, that is important.[104] Beyond these practical issues, I am concerned with the more abstract difficulties the collision of killing and caring produced. And a brief return to the hive might be of help here.

Beekeeping manuals' representation of bees as having the capacity to respond to human intemperance links those small creatures to the other animals with whom people lived, in that we can trace in such representations a sense of the importance of collaboration and of the need for communication across species boundaries with those "poore cattell" as well. What this view of bees as co-workers with humans produces, in turn, is a sense that they were perceived to be not unlike the people who kept them. I am suggesting that we should read this as having both emblematic and practical implications—that the exemplary value of bees emerges from and reinforces the sense that they have experiences that are similar to our own; Joseph Campana has spoken of bees' "strange proximity" to humans in the manuals.[105]

For this reason it is unsurprising that a feeling of empathy and admiration for bees emerges in some manuals.[106] Thus, Lawson argued that bees should not be "impounded" for too long over the winter in part because they were believed to "purge" themselves only outside the hive. "Judge you what it is for any liuing creature not to disburden nature," he wrote. "Being shut vp in calme seasons, lay your eare to the hiue, and you shall heare them yarme and yell, as so many hungred prisoners. Therefore impound not your Bees so profitable, and free a creature."[107] The "yarme and yell, as so many hungred

---

104. In the early eighteenth century, for example, John Houghton recorded the killing of an ox in the following terms: "He is slaughtered by having a Rope put about his Neck and Horns, and drawn through a Hole in a Post, or a ring fastened to it, by which means he is pull'd close, and with a Pole Ax, knock't on the Head, 'till almost or quite dead: Then his throat is cut, and the Blood let out." Houghton, *A Collection for the Improvement of Husbandry* (1692–1703), quoted in MacGregor, *Animal Encounters*, p. 453. Because of their focus on the breeding and raising of (live) animals, agricultural manuals do not describe details of butchery and list, instead, the means to cure and heal animals. Indeed, the techniques used at the time seem rarely to appear in print outside metaphorical usage (which is almost always negative). Tusser, for example, simply notes that "at Hallowtide slaughter-time entereth in, / and then doth the husbandmans feasting begin." Tusser, *Five Hundred Points of Good Husbandry*, p. 44.

105. Joseph Campana, "The Bee and the Sovereign? Political Entomology and the Problem of Scale," *Shakespeare Studies* 41 (2013): 107.

106. Here I am somewhat disagreeing with Vinciane Despret's conception of empathy. She writes that "empathy allows us to talk about what it is to be (like) the other, but does not raise the question 'what it is to be 'with' the other." Empathy is more like 'filling up one self' than taking into account the attunement." Despret, "The Body We Care For: Figures of Anthropo-Zoo-Genesis," *Body & Society* 10, nos. 2–3 (2004): 128. I am suggesting below that the analogies that are used between bees and humans to describe apian behavior in early modern beekeeping manuals reveal a link that goes beyond simile or allegory.

107. Lawson, *A New Orchard and Garden*, p. 24.

prisoners" might be a simile, but Lawson figures their suffering—like their busyness—as natural to, and a recognizable part of the being of, the bees. Likewise, Southerne argued that it was not good to drive bees from their own hive into a new one at midsummer because they would be dejected when they were "forced to gather Waxe and anew to work it, which before in their owne was readie done." He went on: "By naturall reason you may consider, that what man or woman soeuer should be turned out of their lands, goods and houses, the which before they had truly gotten, & should be forced to labour againe, they would rather become desperate."[108] The "naturall reason" that Southerne refers to and that allows him to make this connection between the experience of bees and the experience of humans is understood to be an innate quality that links the species and that allows for fellow feeling. Bees' opinions, Southerne suggests, are ones that people might have, or might (perhaps even should) at the very least imagine having.

A logical extension of this assertion emerges in one of the more self-consciously poetic uses of bees from the period. It was a customary practice of this time to kill the bees in the hive from which the wax and honey were harvested and Shakespeare represented this in a revealing way.[109] Where he had depicted Henry VI likening himself to a cow, so in a later play he had another king linking himself with the bee: however, filial greed and ingratitude, rather than maternal attachment, was this king's focus. "See, sons, what things you are," Henry IV begins:

For this the foolish over-careful fathers
Have broke their sleep with thoughts, their brains with care,
Their bones with industry; . . .
                    when, like the bee
Culling from every flower the virtuous sweets,
Our thighs pack'd with wax, our mouths with honey,
We bring it to the hive; and, like the bees,
Are murdered for our pains.[110]

---

108. Southerne, *A Treatise Concerning the right vse and ordering of Bees*, C4r.

109. Walker and Crane note: "At the end of the summer the bees in the heaviest skeps (i.e. those containing the most honey) were killed, and all the honey was harvested. The same was done with the lightest skeps because they would have insufficient honey to keep the colony alive over the winter. The medium-weight skeps were overwintered, all the honey in them being left for the colonies, and the colonies served as stock for the next year. "History of Beekeeping from c. 1200 to 1850," pp. 232–33.

110. William Shakespeare, *The Second Part of Henry the Fourth*, in *William Shakespeare: The Complete Works*, edited by Stanley Wells and Gary Taylor (Oxford: Clarendon, 1988), 4.3.194, 197–99, 203–7.

Despite referring to it, it is not bees' making of honey and wax that Henry IV is using here. Rather, this royal father's sense of his own labor—his busyness—is underlined by the example they provide. And the king's simile ("like the bees") works because of the established recognition of bees as sharing qualities with humans, because of the potential for empathy across species boundaries. But this is not all that Shakespeare's imagery is doing, of course; and we can turn to the literary theory of the age to further comprehend the full implications of this simile. And the literary theory, in turn, takes us back to the crucial link between emblem and husbandry.

George Puttenham's 1589 definition of metaphor proposes that it is "a kind of wresting of a single word from his own right signification to another not of natural, but yet of some affinity or convenience."[111] "Affinity," from the Latin *"affinitatis,"* also relates to *"affini"*—relatives,[112] and Puttenham's use of the term allows us to see that while the relationship between the vehicle and the tenor—between the image used and the thing described—in a metaphor is not "natural," (that no blood ties are implied, you might say), it suggests proximity. Henry IV's use of "murdered" is usefully read with this definition in mind. It wrests an idea from one realm into another while claiming that those realms have "some affinity." Indeed, it might be said to represent in poetic form the "naturall reason" that Southerne mentioned: in Henry IV's image such reason recognizes that, for the apparently unloved father, as for animals, labor is not rewarded with gratitude but by being reduced to the status of an object that can be dismissed without moral compunction.

Shakespeare's simile thus reflects in one species the uncomfortable double movement of husbandry more generally. A father labors like a bee, so killing a bee is like killing a human. By extension, people labor alongside, in collaboration with, their quick cattle, and the empathetic engagement with animals that is a central and necessary part of husbandry makes troubling the slaughter that is often the aim and end result of it, because that empathy is constituted on the basis of the animals' "strange proximity" to humans. But Shakespeare's image is not to be read simply as an argument against beekeeping. The hive is too productive to be set aside.[113] What the image does is lay bare the paradox at the heart of early modern pastoralism: "killing a

---

111. George Puttenham, *The Arte of English Poesie* (1589), in Brian Vickers, ed., *English Renaissance Literary Criticism* (Clarendon: Oxford, 1999), p. 242.

112. *Oxford English Dictionary*, s.v. "affinity."

113. Indeed, in *The Life of Henry the Fifth*—the next play in Shakespeare's second tetralogy—the Archbishop of Canterbury uses an extended bee metaphor to discourse upon the "natural" order of the human world. Shakespeare, *Life of Henry the Fifth* in *William Shakespeare: The Complete Works*, edited by Stanley Wells and Gary Taylor (Oxford: Clarendon, 1988), 1.2.187–204.

domesticated animal," as McInerney argued, "is tinged with betrayal."[114] The line between slaughter and murder is a fine one, even when it comes to bees.

Recognizing that the paradox of pastoralism was experienced in the fields and yards of early modern Essex has other implications too. Henry IV's use of "murdered" hints that the claim for the existence of a social contract between the species was not in itself sufficient to explain all aspects of husbandry, that while relationships with living animals might be premised upon a conception of reciprocity, killing animals would seem to be doing much more than simply ending that reciprocal agreement. "Naturall reason" tells the king, just as it tells the beekeeper, that killing overwhelms that idea for the very reason that the social contract has reciprocity at its core. The implications are even more troubling when one takes this idea into relationships with cows, which were long term and perhaps more individualized and intimate than with other agricultural animals. If cows were believed to experience special attachment to their calves (which they were) and those calves were understood to be experiencing creatures (which they were), then killing or selling to be killed the calves, let alone the cows, can only be read as marking the utter evacuation of the contract. If you recognize and embrace the fact that it is necessary to "see the Cow stand sure and firme" before you milk her, and—crucially—that it is her decision as to when she does this, then "binding" a calf and "Bearing it to the bloody slaughterhouse" does not fit.[115] Killing cannot be viewed as being part of the social contract of agriculture.

But it is not just cows, those individualized creatures, that witness the failure of the social contract. Likewise, pigs—creatures who would be "in the yard," as some wills note,[116] and so familiar presences as people came to and from the house—were recognized as capable of self-direction. Gervase Markham (refusing the legal understanding of domestic animals) said of them that they were "troublesome, noysome [and] vnruly" but that they were also "the Husbandmans best scauenger."[117] Indeed, it was, perhaps, their very capacity for unruliness that made them such economically valuable creatures.

---

114. McInerney, *The Cattle of the Sun*, p. 36.

115. Gervase Markham, *The English Hovse-Wife* (London: Anne Griffin, 1637), p. 194; William Shakespeare, *The First Part of the Contention of the Two Famous Houses of York and Lancaster* (aka *2 Henry VI*), in *William Shakespeare: The Complete Works*, edited by Stanley Wells and Gary Taylor (Oxford: Clarendon, 1988), 3.1.211–12.

116. ERO, D/ABW 46/74 (1625), made probate 72 after writing; ERO, D/ACW 10/243 (1628), made probate 319 days after writing.

117. Markham, *Cheape and Good Husbandry*, p. 123.

Thus, despite the fact that, in Markham's terms, they were "excellent obseruers of their owne homes," Josselin recorded that on 5 July 1650,

> I was troubled with my hogs breaking away on the lords day morning, and having looked for him and not finding him. I observed still a vexacon and trouble in all thes things and thinking gods providence might bring him backe to mee, I was just then told that he was driven home from of the greene.[118]

The "trouble" Josselin records is the annoyance caused by his pig's wandering, but that very wandering might be how such an animal found food and so might be why it was regarded as "the Husbandmans best scauenger." (What is also notable in Josselin's diary entry is the fact that it links husbandry and religion, as we have come to expect.) Slaughtering such animals was thus removing what were recognized as self-willing agents from the household. And, as the example of bees showed, the fact that they were not unlike humans might make it inevitable that killing them could "tinged with betrayal" and might, poetically, be seen as murder. Only the slaughter of sick animals or those who were felt to be in pain could possibly be viewed as being for the benefit of the animal: no healthy calf or pig was ever likely to benefit from its own death. And killing an aged cow because she was no longer productive is likewise difficult to see as being for her good, because her good might exceed her productivity. Thus, slaughter—which was so central to maintaining a productive herd and to the well-being of the human family—must almost always be regarded as being done with only human interest in mind, and the idea of a social contract existing between humans and animals becomes as difficult to see in action in preindustrial as it is in industrial agriculture. As the environmental ethicist Clare Palmer puts it: "The language of the animal contract—implying as it does, some kind of consent or agreement—serves to legitimate the power which humans have acquired over domesticated animals."[119]

If the social contract cannot explain slaughter on the farms and smallholdings of early modern Essex, perhaps the concept of humanity's God-given power over the animals might be more helpful. The idea of dominion was, of course, decreed in the Bible. Genesis 1:28 reads: "And God blessed [Adam and Eve], and God said unto them, Be fruitful, and multiply, and replenish the earth, and subdue it: and have dominion over the fish of the sea, and over the fowl of the air, and over every living thing that moveth upon the earth." In

---

118. Josselin, *The Diary of Ralph Josselin*, p. 209.

119. Clare Palmer, "The Idea of the Domesticated Animal Contract," *Environmental Values* 6, no. 4 (1997): 422.

theory, therefore, things were clear: "The Creatures were not made for them-selues," wrote Henry Vesey in 1621, "but for the seruice and vse of Man."[120] In practice, however, things were a bit more complicated. When it is recog-nized that the "seruice and vse" of animals included the need for collegiality across species barriers and involved the acknowledgment of and, more, reli-ance on the fact that animals themselves could also experience experiences, such a claim for the straightforward human right to exercise power over the so-called lower creatures must be recognized as being somewhat less than helpful. The animals people worked with—with their moral uprightness, maternal instincts, and unruliness—were clearly also "made for themselues," to use Vesey's words, and utilizing their being "made for themselues" was actually part of the process of good husbandry. In the yard, therefore, the theory of dominion would have been present, but it would have had to sit alongside the sense of one-to-one, personal engagement between people and their quick cattle. And the significance of the personal engagement cannot be dismissed.

To ask how it was possible that the testators of early seventeenth-century Essex could look a pig in the eye and knock it on the head with a pole axe is not, therefore, to force onto early modern culture a modern sentiment that emerges out of the distance between consumer and meat animal that many people can take for granted in the twenty-first century. Rather, I think the question is appropriate because the difficulty of living with and killing animals was an unavoidable product of the closeness that early modern people expe-rienced in their relationships with them, and was inevitable given their belief that the animals themselves experienced those relationships too. Indeed, to turn it around, if we assume that early modern people found slaughtering their animals unproblematic, we would seem, by implication, to be anesthe-tizing those people, making them incapable of feeling. This, I think, would be misreading the past. And yet, of course, they killed their animals.

John Berger offers a way of thinking about this issue that brings us back again to the role of the emblematic worldview in the fields and yards of early modern Essex. Writing about twentieth-century rural culture, he proposed that "a peasant becomes fond of his pig and is glad to salt away its pork," arguing that "what is significant, and so difficult for the urban stranger to understand, is that the two statements in that sentence are connected by an *and* and not by a *but*."[121] I am suggesting that for early modern people, too,

---

120. Henry Vesey, *The scope of the scripture* (London: W. Iones, 1621), p. 8.
121. John Berger, *About Looking* (London: Writers and Readers Publishing Co-operative, 1980), p. 5.

those statements were joined by an and and not by a but and that that "and" that existed in the early seventeenth century was available because people had access to a particular realm of meaning that allowed an animal's death to have more than practical significance. My wording—"and that that 'and' that existed"—is deliberately messy here because the idea is so remote from an accepted, twenty-first-century sense of categorical clarity because theirs was an "and" that has faded from view: as empirical science has displaced the emblematic worldview as an explanatory device; as agricultural animals have come to be more and more easily regarded as creatures with only material meaning for their consumers.[122] And it is the very juxtaposition of conversation and killing in early modern husbandry that affords the final—perhaps the ultimate—role for the emblematic worldview in early modern England. The "and" that links fondness and killing is the symbolic potential of animals.

Comprehending animals' utility as existing beyond their fertility, labor, financial value, and flesh made the complexity of people's relationships with them more bearable because that complexity was experienced as being meaningful. Indeed, it is possible that it was this early modern sense of animals as possessing both material and immaterial value that took up the role that sacrifice had played in ancient Greek culture. The emblematic worldview does not, of course, make the discomfort of the juxtaposition of collaboration and slaughter disappear. Rather, it values that discomfort, it gives it a moral meaning. Reading slaughter as part of the "living language of the Creator" might seem paradoxical, but it makes the difficulty of killing consequential in worldly and otherworldly terms. And, as is visible in the bequest of a lamb to a godchild, the symbolic and the real are, once again, understood to be co-constitutive. Indeed, the gift of a lamb had financial and practical value, and was also an emblem of salvation, only because the animal would at some point in its future be slaughtered.[123] Without the possibility of its death, the full value of the bequest would be lost. To put it another way, a lamb is meaningful in early modern England because it is, in current parlance, always already killable.[124]

---

122. I return in the afterword to the problem of the "and" for producers in current agriculture.

123. The only exception would be if human control was lost—that is, if illness or a predator intervened to kill the lamb.

124. See Donna J. Haraway, *When Species Meet* (Minneapolis: University of Minnesota Press, 2008), p. 80. It is worth noting that some animals seem to carry more emblematic meaning than others; and so—by logic—their deaths might be more easily experienced. The beekeeping manuals make it clear that these insects had vast symbolic potential, and lambs possessed a link to Christ and so gained meaning from that. But cows do not have such apparently immaterial qualities. Topsell's reading of them, for example, lacks the kind of scriptural associations that he is able to make with lambs, and the fact that they seemed to have less to say "in the language of the Creator" may have made worse

It should come as little surprise, therefore, that some Protestant thinkers in the early seventeenth century went further and recognized this otherworldly value not only in the living creatures or in the killing of them but in their dead bodies too. Consuming animal flesh was also presented as possessing emblematic meaning. Just as Heresbach, a Calvinist, argued that bees produced honey, wax, and examples, so John Paget, a Reformed English minister living in exile in the Netherlands,[125] presented meat eating in uncompromising and yet utterly emblematic terms:

> Our food is not onely of corruption, but we feed even of death it selfe, & that by the allowance of God, *Gen.9.3* in taking away the life of other creatures to maintaine our owne; especially in these last times when he hath said unto us of them all, Rise, kill & eat, *Act.10.13.* Whatsoever is sold in the shābles, that eat, asking no questiō for conscience sake. *I Cor.10.25.* herein we see death dayly presēted to us & set before us on our tables. This is seriously to be thought upō as a wonderfull work of God: by the death of other creatures our life is preserved: our living bodies are sustaind by their dead carcasses: in their blood swimmes our life, and from their pangs of death spring the pleasures of our life, our feasts & ordinary food.[126]

These animals sustain us, but not in the way that Thomas Dekker's North-Country-man put it. He wrote of quick cattle as "our faithfull Seruants . . . our Nurses that giue vs Milke . . . our Guides in our Jornies . . . our Partners."[127] Paget proposes that animals sustain us more literally: "in their blood swimmes our life." Not only always already killable, animals are also always already placed in a strange proximity to humans, even when human power over them would appear to have been exercised with such finality. And so for Paget, the troubling closeness of animals is core to the meaning of their flesh, and to become a vegetarian as a response to this closeness would be to stop reading

---

the discomfort experienced by the people they lived alongside when they sent them or their young to slaughter. Perhaps it was this symbolic lack that emphasized the role of the cow in this rather than the other world. A lamb's obedience to its dam's voice makes it like Christ; a cow mourning its lost calf makes it like a human. That is troubling. Distancing oneself from the slaughter of these animals might have been particularly important.

125. Keith L. Sprunger, "Paget, John (d. 1638)," in *Oxford Dictionary of National Biography* (Oxford: Oxford University Press, 2004).

126. John Paget, *Meditations of Death Wherein a Christian Is Taught How to Remember and Prepare for His Latter End* (Dort: Henry Ash, 1639), p. 49.

127. Thomas Dekker, *The cold year. 1614. A deepe snow: in which men and cattell haue perished, to the generall losse of farmers, grasiers, husbandmen, and all sorts of people in the countrie; and no lesse hurtfull to citizens* (London: W. W., 1615), B2v.

the "living language of the Creator" as it was represented daily in the killing of his creatures, which would, in turn, make animals' lives and deaths meaningless.[128] It would be a turning away from something that was, like God's provision of pelicans, bees, and lambs, a "wonderfull" work of the divine will. Meat was there to feed the mind as well as the body.

But Paget goes further and notes that the social separation that might allow the rich and the urban to be at a distance from slaughter does not offer freedom from the discomfort that eating animals can and should produce. Indeed, where the husbandman might, as Herbert knew, make great use of knowledge of "pastorage" in understanding God's universe, Paget asked those who were distant from killing to engage in what might be a more active kind of contemplation in order to achieve similar knowledge of the divine.

> The Gentleman that sits at his table above in his dining chamber, and was not present in the kitchin or butchery, to see the execution, the convulsions of death, the sprinting & gasping of the slaughtered creatures, is yet by remembrance to represent the same and to make it present againe in his eating: for eating & burying of them in our bellies is more then killing of them, & a further meanes to strike the heart with thought of death, procured for the eater.[129]

How a gentleman might "by remembrance" make the butchery present is not made clear, but perhaps the implication is that everyone in this period would at some time have seen an animal being killed and that that experience should be brought to mind at the meal table. Or perhaps it is that the very language of Paget's text itself (its emphasis on "the sprinting & gasping of the slaughtered creatures") works—like a good sermon should—to bring the killing vividly to life.[130] But the "remembrance" might also—to return to the

---

128. See Erica Fudge, "You Are What You Eat," *History Today* 67, no. 2 (February 2017): 41–46. The value placed on the difficulty that eating meat produces fits with the Protestant argument that the struggle with the world and the worldly is the task of the good man, and that avoiding the world and the flesh would be a failure to exercise virtue. Joseph Hall wrote, for example: "I would not bee a Stoïcke, to haue no Passions: for that were to ouerthrowe this inward gouernment God hath erected in me; but a Christian, to order those I haue." Hall, *Meditations and Vowes Diuine and Morall* (London: Humfrey Lownes, 1607), Book I, pp. 98–99.

129. Paget, *Meditations of Death*, p. 52.

130. In the prefatory note "The Publisher to the Reader" in *Meditations of Death*, Paget's nephew Robert Paget writes that "it containes the summe of that which was delivered in divers sermõs to his owne flock in the year 1628." Paget, *Meditations of Death*, 5r. It is possible that Paget was influenced by Plutarch's *De esu carnium*, which was translated into English by Philemon Holland in 1603. In that translation Plutarch asks of the man who first "approached with his mouth unto a slaine creature": "how came it that his taste was not cleane marred and overthrowen with horrour, when he came

beekeeping manuals—be an act that calls up "naturall reason," an act that requires an empathetic contemplation of all that we share with animals.[131] The call to represent slaughter to oneself at the dinner table should be read as a call to imagine one's own death and one's own edible fleshiness, and thus eating meat is a memento mori. For not only are we in death in the midst of life, we are also all, as that ungrateful son Prince Hal knew, food for worms.[132] And eating animal flesh was understood as an important reminder of this.

Paget was not alone in his reading of meat as a memento mori: John Moore, another clergyman, also saw meat as having a vital place in the Christian's contemplation of their worldly and otherworldly being, and I quote at length from his 1617 text *A Mappe of Mans Mortalitie* because the journey Moore takes from humanity's mortality and absolute materiality to our species' need to contemplate the eternal through animals resonates so fully with the issues this chapter has been concerned with:

> Our life is as a garment that weares of it selfe, and by itselfe; for we weare out our life in liuing; the more we liue, the lesse we haue to liue, and still approach nearer death: whatsoeuer we are cloathed with, is a mortall and perishing merchandise; our garments weare vpon our backs, and we in our garments; they are eaten with mothes, and we with time. So in our meates (as in a looking-glasse) we may learne our owne mortalitie: for let vs put our hand into the dish, and what doe we take, but the foode of a dead thing, which is either the flesh of beasts, or of birds, or of fishes, with which foode wee so long fill our bodies, vntill they themselues be meate for wormes? All this we see by experience, we feele it and we taste it daily: we see death (as it were) before our eyes:

---

to handle those uncouth sores and ulcers; or receive bloud and humours, issuing out of the deadly wounds." Plutarch, "Whether it be Lawfull to Eat Flesh or No," in *The Philosophie, commonlie called, The Morals*, p.572.

131. This sense of affinity was expressed in a particular way in relation to pigs. Topsell noted: "there is so great resemblance or similitude betwixt a mans flesh and Swines flesh, which some haue proued in tast, for that they haue eaten of both at one Table, and could find no difference in one from the other." Perhaps it is this that gives emblematic value to animals that are otherwise lacking in positive symbolism. Indeed, Topsell commented on hogs' cannibalism, on their "abstain[ing] not from fat Bacon," a habit which, he suggested, distinguished them from other animals: "Dogges will not taste of Dogges flesh, and Beares of Beares, yet will Hogges eat of Swines flesh." And he goes further: "yea many times the damme eateth hir younge ones" as if to explicitly differentiate pigs from cows whose maternal feelings were so central to understanding and working with them. Topsell, *Historie of Foure-Footed Beastes*, pp. 678 and 667.

132. William Shakespeare, *The History of Henry the Fourth*, in *William Shakespeare: The Complete Works*, edited by Stanley Wells and Gary Taylor (Oxford: Clarendon, 1988), 5.4.86.

we feele it betwixt our teeth, and yet can wee not cast our accompt, that we must die.[133]

"(As it were)": this, like Paget's gentleman's remembrance of slaughter, is an act of imagination, but is still potent for all that. Moore continues, noting humanity's continued failure, as Philagathus put it, to "leaue off this talking of . . . worldly matters."

> Many make a couenant with Death, and clap hands with the graue, hoping thereby to escape, or for a time to solace themselues in the forgetfulnesse of their latter end, and so bathe themselues in their fleshy pleasures, and wallow (like fatted Swine in the filthy stie) of all vncleanenesse, still following things apparant to their eyes, and neuer regarding the time to come, till death preuēt them on a sodaine, and summon them to appeare before their Iudge.[134]

Humans are just like all the other stuff in the world: like clothes, we wear out. Writing a will, as the Essex testators understood, was a recognition of this; it was an act by mortal beings deliberating about material stuff. But what this chapter's reading of a number of the specific animal bequests reveals is that some of the animals that were bequeathed in those wills might actually take testators outside of the realm of the flesh and into the realm of the spirit, and it is here that we can see how killing animals, like contemplating the hive, like bequeathing a lamb, carried both worldly and otherworldly resonances.

The social contract can only take us so far in understanding how testators and their legatees understood their animals, then. Indeed, its attempt to rationalize killing—to make slaughter part of a contractual agreement—works against some ideas that can be traced in the early modern period: about people's recognition of animals' potential for recalcitrance, of the acknowledgment of cows' maternal instincts and bees' moderation. In short, instead of rationalization, what we find in the yards and fields of Essex is the work of empathy, and thus experiencing discomfort at the death of one's animals should be understood as a core part of their function.

So, to return to a will we have already encountered, when Mary Archer, a grandmother from Little Clacton, requested that "myne executor tak in my two old cowes at Micholmas that are now in the possession of Henrie Soyars and that he shall sell them and putt out two young Cowes in sted of them to

---

133. John Moore, *A Mappe of Mans Mortalitie* (London: George Edwards, 1617), pp. 39–40.

134. Moore, *A Mappe of Mans Mortalitie*, p. 42.

the vse and benefitt of John Archer and Robbecah Archer the two youngest children of my sonne Thomas Archer,"[135] we can now wonder if she was doing more than simply condemning her old cows to slaughter to provide a future for the younger generation. It is likely that these animals—perhaps being looked after by Soyars, a neighbor during her final illness[136]—had lived with Archer for years, had provided her household with nutrition and wealth, and had been important daily presences in her life. The bequest signals that her grandchildren's futures were more important to her than her old cows, undoubtedly; but that does not mean that the cows were not also significant; that they were not made for themselves. Her will, perhaps marks the end of their relationship in more than practical terms. It might be that what is evidenced in Mary Archer's bequest is the final acknowledgement of their shared mortality. Archer may have been "pfect in Memorie" when she dictated her will to her scribe, as the law demanded, but she was also, perhaps, using her "naturall reason"—a capacity that exceeded legal requirements—to recognize a link between herself and the beasts whose calves had been a source of income to her for years. The bequest may have been her admission, in fact, that, like her, her "two old cowes" were also grandmothers and that she was herself just aged flesh.

Such a reading enables us to see that early modern husbandry should not simply be read as the straightforward enactment of human dominion— the exercise of the belief that quick cattle were created only and simply for humans to use. It reveals the possibility that people viewed their existence alongside animals as part of an entwined system of practicality, partnership, and symbolism. Animals, in this context, were simultaneously collaborators in husbandry and extraordinary beings constructed by the Almighty for wonder and for use. Bequeathing a hive for a sermon, leaving a lamb for a godchild: what was happening in both of these instances announces what might seem utterly paradoxical to us—that the worldly matters of the field were also replete with otherworldly value. This value helped to support a system that constantly breached the contract humans had made with the animal co-workers with whom many people lived side by side. There was always a hierarchy, of course; but it was a hierarchy that did more than simply allow for, but also actually relied on the tangled emblematic and practical complexity of life. It was a hierarchy that knew animals were responding and not just

---

135. ERO, D/ABW 43/197 (1620), made probate 107 days after writing.

136. Other wills seem to witness this kind of care for animals by family or neighbors. See, for example, ERO, D/ABW 43/71 (1621), made probate 25 days after writing; ERO, D/ABW 46/160 (1625), made probate 140 days after writing; and ERO, D/AEW 19/141 (1632), made probate 101 days after writing.

reacting to human wishes, that they were capable of forming relationships among themselves that mattered to them, and that they were "living characters in the language of the Creator." In this way animals were worldly as well as otherworldly matter, and that, in turn, required killing them to have, however uncomfortably, a meaning. Indeed, the discomfort experienced in that event was a crucial part of its meaning. What happens to such a worldview when a separation is made between conversation and killing, when households cease to include animals but continue to encounter slaughter, is a focus of the next chapter.

CHAPTER 5

Less than Kind

The Transient Animals of Early Modern London

The will of Thomas Stevens, a husbandman from Ulting, which was written on 13 May 1626, contains eight bequests: 5s to his brother John, 10s to his unnamed sister, 40s each to Ellen Hodge and Susan Kitto, 20s to his kinswoman Roofe Barbor, 5s to Annis Uskum, 3s to Elizabeth Goodman, and "all my waringe apparel" to Jolian Writter. The document is witnessed with John Alexander's signature and John Pratte's mark. So far, the will follows convention. But it does not end there, as would be customary. Instead, the scribe—the same one who had written the will—added another paragraph reporting the following:

> The sayd Thomas Stevenes was earnestly vrged by theis men heare vnder written John Wiseman John Alexander and John Pratt to giue the greatest parte of his goods vnto his brother and his Sister but his answer was that the he hade abewsed him and kicked him Lycke adodge a dogge and theirfore he wood not bestowe any more of him.

Alexander and Pratte witnessed the document once again, and that is its conclusion. What follows is the usual note recording the place and date of probate added by a legal clerk (Kelvedon, 27 September 1626), with an interesting addition—a confirmation in that same clerk's hand that Ellen Hodge and Susan Kitto were legatees.[1] This signals, perhaps, that Stevens's brother had attempted to contest the will and its bequeathing of more money to those two women than to him. The confirmation shows that the testator's wishes prevailed, however, and that the abusive sibling received what Thomas intended.

Stevens's will is unusual in a number of ways. Not only is the note added by the scribe exceptional and not only does the legal clerk usually only report the place and date of probate, these documents rarely include such details of family breakdown. Instead wills typically offer glimpses of functioning kinship and neighborhood networks. Their distribution of property reveals family ties, friendships, responsibilities being fulfilled: a debt to be paid, a helpful neighbor to be recognized, a godchild to be remembered. Just as they attempt to ensure the smooth continuation of the world the testator is about to leave—with households, husbandry practices, small businesses to be carried on after them—so they reveal affection, gratitude, care. But every now and then a testator used the power that writing a will confers to say, and so to reveal, something more, and in so doing they expose family life in a way that reminds us that behind the conventional words of wills lie real people with potentially complicated domestic worlds.[2]

Thomas Stevens's will achieves this exposure brilliantly, and the language used in his brief declaration about his brother—that he "hade abewsed him and kicked him Lycke . . . a dogge"—is particularly telling. He links a breach of kinship with unnatural species descent: his brother's viciousness represented as having undermined his (Thomas's) humanity. Indeed, Thomas's pairing (in its absence) of family feeling with being human is an echo of a more poetic work from about a quarter of a century earlier that affords a useful way of thinking about the core issue at stake in this chapter. First performed in the Globe Theatre on London's Southbank, about thirty-five miles away from Stevens's home in Ulting, Shakespeare's *Hamlet* has the eponymous prince describe King Claudius as "a little more than kin and less

---

1. ERO D/ACW 10/214 (1626), made probate 137 days after speaking.

2. Robert Wyles, a merchant from Dedham, attempted to make his daughters take their mother's advice by threatening to halve their legacies if they ignored her; ERO, D/ABW 47/162 (1625), made probate 100 days after writing. Christopher Steell, a yeoman of Great Totham, tried to control his wife Joane and her "turbulent humor" by threatening to replace her generous legacy with 12d if she asked for more; ERO, D/ACW 9/229 (1625), made probate 36 days after writing. John Birch, a husbandman from Hatfield Peverel, referred to his "Vnkinde Wiffe," leaving her only 1s 6d because of the "trouble & greefe" she had caused him; ERO D/ACW 11/21 (1629), made probate 54 days after writing.

than kind."[3] Now married to his brother's widow, Claudius is both Hamlet's uncle and his stepfather: he is family in an excessive way ("a little more than kin") and he is "less than kind"—simultaneously ungenerous in his lack of emotional care and not quite human. Where Stevens's story linked the breakdown of the family with violence and to a lowering of his own species status, Hamlet connects a breakdown of the family with (legal, moral, emotional and, ultimately, physical) violence and to the loss of a sense that he shares a nature with his uncle. The double meaning of "kind" in Hamlet's speech—its compressing into one word the ideas of generosity and of community—draws our attention to what is, perhaps, only implied in Stevens's will in which John Stevens's being unkind un-kinded his brother—turned him into a lesser being. This double meaning of "kind" makes care a product of identification, makes being kind a matter, you might say, of being of a kind. It may be coincidental that Stevens and Shakespeare both represent broken family ties in these terms, but the accident is productive, and this chapter takes up the double meaning of the term "kindness" that Hamlet uses to think about human-animal relations, using it to consider what happens to empathy for quick cattle when animals are not present in the life of the community in the same way as they had been; how kindness—generosity and recognition—was enacted in changed circumstances.

Thus, instead of looking at the meanings and implications of the existence of animals in the lives of people, as has been the focus of the book so far, this chapter will speculate about what it might have meant to those people to live without quick cattle and how that might have affected the way animals were perceived and treated. This might sound like a historical thought experiment rather than an analysis based on the interpretation of data, but it is, in fact, a response to a real fact of life in the period studied here. The growth of London impacted not only the people who migrated there but also the countryside around it, and human relationships with quick cattle were among the key aspects of life that were affected. The chapter will track these changes by comparing the wills written in Essex with those written and made probate in the Diocese of London in the period 1620 to 1635.[4] What the London Metropolitan Archive (LMA) dataset reveals is that remarkably few testators from

---

3. William Shakespeare, *Hamlet*, in *William Shakespeare: The Complete Works*, edited by Stanley Wells and Gary Taylor (Oxford: Clarendon, 1988), 1.2.65.

4. "Up until 1845 the [Diocese of London was] comprised of most parishes in Middlesex . . . [and] the City of London parishes. . . . [In addition,] the diocese retained nine Essex parishes: Barking, Chingford, East and West Ham, Little Ilford, Low Leyton, Walthamstow, Wanstead and Woodford." "Diocese of London," National Archives, http://discovery.nationalarchives.gov.uk/details/rd/ee256443-9b57-4f3b-ba89-4971b67ded5b, accessed 11 October 2016.

the capital included specific animal bequests in their wills, and attempting to understand what that lack of record might signify about closeness and compassion—about kindness—will be the focus here.

Thomas Stevens's declaration that his brother "hade abewsed him and kicked him Lycke . . . a dogge" also helps us think about the issue of lack of documentary record. It shows us how complex the nature of historical evidence can be. Without that additional, unusual, unnecessary note, the story of Stevens's family life would never have been visible to us, something that should serve to remind us that a will's distribution of property can, on its own, make available only some of the world of the testator. Who gets what (a lamb, for example) is important, but so is why someone doesn't get anything at all—but the latter might be impossible to trace: not leaving something to someone in a will is also not leaving anything for the historian to interpret.

The reasons that underpinned the choices testators made are usually absent. But that, of course, does not mean that any hope of interpretation should be abandoned. As Hannu Salmi has noted, "When the past is silent, the historian must also consider the ways in which information can be gleaned about the invisible . . . must use extensive materials, compare data, and, ultimately with the aid of a comprehensive interpretation, make deductions about the object."[5] This chapter will try to make deductions about the lives with and of quick cattle in early seventeenth-century London. It will attempt through the new dataset of wills and through the picture of urban life other historians have constructed a comprehensive interpretation of their world. The chapter begins, however, not in the capital but in more familiar territory forty-nine miles away,[6] with one extended family in Essex that provides a foil for what follows. Theirs is an animal-filled world, one in which quick cattle, to invert Hamlet's image, might be described not as "a little more than kin and less than kind" but rather as "a little less than kin and more than kind." Not related to the humans they lived with, but affined to them, animals were understood in this world to be generous in their provision to the people who lived with them and were simultaneously recognized as something more than chattels. It is this world, I suggest, that is absent in the capital.

Just three years before her neighbor Christopher Fuller married Ellis Myll in the same church, on 4 July 1586, Margery Cossen married Thomas Reade

---

5. Hannu Salmi, "Cultural History, the Possible, and the Principle of Plenitude," *History and Theory* 50, no. 2 (2011): 175.

6. Throughout this chapter I have taken St Paul's Churchyard as the measuring point for distances from London, using the calculating tool at "As the Crow Flies" Distance Calculator: http://tjpeiffer.com/crowflies.html.

in St Mary the Virgin in West Bergholt.[7] Over the following years Margery and Thomas, a yeoman, had three sons (Thomas, John, and William) and six daughters (Marie, Elis, Elizabeth, Eme, Margery, and Jane) baptized in that church. Thomas the elder was buried there on 20 May 1617, just four days after the minister, Gregory Holland, wrote his will for him. Two daughters— Elis and Eme—were not mentioned in their father's will, so it is likely they had died before it was written. The will contained over £300 in cash bequests to his family (which included two grandchildren) and, as well as kitchen equipment, beds and bedding, and other furniture, Thomas senior included a bequest to Margery his wife of "sixe of my best bease a Ewe sheepe, one nagge or mare at her choyce" and "iiii milke boules my lesser cheese [illeg.], one cherne ^3^ cheese motts [molds] & breads [boards]."[8] He left his real estate, which included a "farm house called the Amry wth all the Meads & lands," to his oldest son Thomas; and four years later, in the survey taken in the village, Thomas the younger, was listed as having over 133 acres of leasehold, and was the second largest tenant on Sir John Denham's land in West Bergholt.[9]

Eight and a half years after her husband's death, on 26 November 1625, Gregory Holland wrote Margery's will for her, just four days before her burial. She signed with a mark, just as her husband had done. In her will, Margery revealed herself to still be in possession of at least £13 of the £30 her husband had left her, a "litle Cowe," a "blacke Cowe," and another cow, a "graye mare," and four sheep and five lambs.[10] Like her possessions, her family had changed somewhat as well. In addition to the three sons and two daughters she mentioned (Marie and Margery are not included in her will, which suggests that they had died between 1617 and 1625[11]), she bequeathed items to one son-in-law, one daughter-in-law, eleven grandchildren, and one godchild whose own child also received a legacy. The youngest of her grandchildren, Rose Reade, the daughter of Thomas and his wife Rose, was only eleven

---

7. Parish Register of St Mary the Virgin, West Bergholt, ERO, D/P 59/1/1 (1559–1658). Margery and Thomas Reade's daughter Jane married Jeremy Mills, and it is possible—given the nature of early modern spelling—that this Mills was related to Fuller's wife.

8. ERO, D/ACW 8/31 (1617), made probate 15 days after writing. The motts and breads were cheese-making equipment. *Oxford English Dictionary*, s.v. "motts" and "breads."

9. ERO, D/ACW 8/31; "A survey taken in September last 1621of the lands of Sr John Denham knight," ERO, D/DMa M18 (1621).

10. ERO, D/ACW 10/76 (1625), made probate 21 days after writing. Margery Reade possessed "at least" £13.

11. There may be another reason for Marie's exclusion from her mother's will. The parish register of West Bergholt records the baptism of Margaret, "the base [i.e., bastard] daughter of Mary Reade" on 27 December 1615; see ERO, D/P 59/1/1.

months old when Margery died. The fact that the baptisms of three of her grandchildren (a child of Jane and two children of Thomas) and her daughter Elizabeth's marriage were also included in the West Bergholt parish register signals that at least some of the family remained local. Indeed, Abraham Barrell, the son of her late daughter Margery, was apparently living with his grandmother at the time of her death because she left Jeremy Mills her son-in-law £4 "vpon the condicon yt the sd Jerem mills shall take Abraham Barrell my grand child immediatly after my decease & shall keepe him . . . till he be twenty yeares of age."[12] Margery Reade's care for her family can thus be seen in her will in two ways: in her children's survival and fertility—the size of her clan marks their health, which was her maternal responsibility—and in her painstaking sharing of her possessions among them. Everyone got something and the guardianship (and so the safekeeping) of one grandchild was ensured. Her will was thus a method of securing the family's future in more than just material terms.

But there is also another way that Reade's will reveals care—a way that studies of early modern charity have yet to take fully into account. While the holding of church ales to help a neighbor to raise money to buy a new cow has been recognized, widow Reade's will records a different kind of support being offered. Her bequest of a cow to her thirty-year old daughter Elizabeth specifies that the latter would receive "my blacke Cowe wche I lent her this last year."[13] It is impossible to say how common such loans of cows might have been in this period (such transactions would usually have taken place without documentation), but other wills seem to reveal the practice too. On a purely practical basis, such loans made good sense:[14] they might be made to attend to an immediate need created by the death or illness of the borrowing household's own animal; a need that would be important to address as a cow was so continuously significant to the family's well-being. These loans were a means by which domestic stability might be restored and maintained. For this reason, the lending out of a cow might

---

12. ERO, D/ACW 10/76. The fact that Abraham was Margery junior's son is made clear in Thomas senior's will.

13. ERO, D/ACW 10/76.

14. John Benton, a yeoman from North Weald Bassett, left his daughter Susan "three milch beastes wch now she hath in hir occupying"; ERO, D/AEW 18/78 (1626), made probate 112 days after writing. Luke Hockly, a husbandman from Great Warley, bequeathed his brother-in-law Matthew Hogate "on cow now in his own hands"; ERO, D/ABW 46/39 (1625), made probate 539 days after writing. Other wills refer to the "letting" of animals—perhaps a financial rather than a familial transaction. See ERO, D/ABW 45/183 (1623), made probate 69 days after writing; ERO, D/ABW 46/190 (1624), made probate 23 days after writing; ERO, D/AEW 19/180 (1633), made probate 33 days after writing; and ERO, D/ABW 50/17 (1630), made probate 84 days after writing.

be said to reveal their more-than-material significance in a way related to discussions in the previous chapter. They were a means by which family ties were reinforced.

But lending a cow was giving more than a source of milk; it was also a mark of an attentive understanding of the animal. This might seem counterintuitive, as loaning a cow could be read as marking the animal as simply a movable thing—a piece of property that could be shunted around. But a cow's place as a possession that might be loaned out should not be read as a cancellation of the close relationship her owner might have had with her. The loan, in fact, can be read as illustrating the knowledge of the specific animal being offered. No mother would lend their daughter a cow who would not stand to be milked, who would not display gentleness. As Gervase Markham put it, showing the double meaning of kind, once again, a good cow "must be kind in her own nature; that is, apt to conceive, & bring forth, fruitfull to nourish, and loving to that which springs from her."[15] Because she was stuff that could be given, you might say, this did not stop a cow from being understood as an individual who could give.

But we can read even more into the loaning of cows than this. Historians have recognized that the concept "family" was different in early modern culture. Ann Kussmaul, for example, has suggested that servants as well as "all those . . . who lived under the authority of the *pater familias*" would be understood to be part of the family in this period, not just individuals related by blood and marriage—and wills in the large ERO dataset seem to reflect this, with legacies for servants recorded in a number of documents.[16] Indeed, Naomi Tadmor has gone further than Kussmaul and has proposed that we should use the term "household-family" rather than simply "family" to reflect more accurately the broader conception of who would have been included back then. And it is notable in the context of this book that Tadmor presents the nature of engagements among people in such household-families

---

15. Gervase Markham, *The English Hovse-Wife* (London: Anne Griffin, 1637), p. 193.

16. Ann Kussmaul, *Servants in Husbandry in Early Modern England* (Cambridge: Cambridge University Press, 1981), p. 7. Sheep and lambs were the most common animals to be left to servants. Twenty-seven servants, maids, or apprentices were left animals in the large ERO dataset. One will that bequeathed a sheep to "eu'y of my servants" is excluded from this total because of the ambiguity about the numbers referred to. Twenty of the twenty-seven bequests (74 percent) were of sheep and/or lambs, five (19 percent) were of cows, one was of cows and sheep (3.5 percent), and one was of horses. Forty-four percent of the twenty-seven servants were female. When a bequest was made to an individual without naming them as a servant and no family relationship was stated, I placed that individual in the category "neighbor." It is not possible to know how many such "neighbors" were actually servants (or even family members). The will that left sheep to "eu'y of my servants" is ERO, D/ABW 47/195 (1625), made probate 90 days after writing.

in terms that are familiar to thinking about human-animal relations. Just as Bernard E. Rollin has proposed that the social contract of husbandry was premised on the exchange of labor for care, Tadmor writes that in a world where adolescents and young adults from outside the immediate family often lived and worked within the household, "household-family contracts involved an exchange of work and material benefits."[17] These human-human contracts did not end with the almost inevitable slaughter of one of the parties, of course, but Tadmor describes the experience of them in a way that resonates with thinking about relations between people and their animals—or rather, that allows us to see that the line between human-human relationships and human-livestock ones might not have been so clear then as seems to be the case now. For some, Tadmor writes, the household-family "is best understood in institutional and instrumental rather than sentimental terms . . . [but] this is not to say that emotions were unimportant in the context of household-family. . . . Instrumentality and affection often went hand in hand; indeed, an increase in affection often led to an increase in instrumentality."[18] Where servants might have this dual role—be useful, and be cared for—so quick cattle, creatures, who were simultaneously of emotional and instrumental value—who were agents and things, symbols and goods, colleagues and meat likewise made many households function, and I wonder if it is possible to suggest that animals were also considered to be part of the household-family in a way that is now lost. The wills reveal that cows in particular acted not only as instruments of care but also as objects of care—and this reflects their being integral to the household's well-being. They were essential to it: that is, they were necessary but they were also intrinsic. It is in such a context that the poor man's desperation over the loss of his cow in Thomas Lodge and Robert Greene's *A Looking Glass for London and England* might best be understood.[19] And the link between human and animal families is reflected nicely in one will, written in 1630 but made probate too late to be included in the large ERO dataset. In it the yeoman John Nysum of West Mersea bequeathed to his sister Rose "the Redd Cowe wth the white face" and to another sister Margaret "the bullock which came of Roses Cowe."[20]

---

17. Naomi Tadmor, "The Concept of Household-Family in Eighteenth-Century England," *Past and Present* 151 (1996): 124. Here Tadmor is using evidence from the mid-eighteenth-century shopkeeper Thomas Turner's records of his relationships with members of his household, but her argument is that his attitude was representative of the longer period.

18. Tadmor, "The Concept of Household-Family in Eighteenth-Century England," p. 124.

19. Thomas Lodge and Robert Greene, *A Looking Glass for London and England* (London: Barnard Alsop, 1617), B4r.

20. ERO, D/ABW 52/267 (1635), made probate 1,744 days after writing.

Like the loaning of a cow to a daughter, the circulating of and cohabitation with animals offers a glimpse of a key way of thinking about the place of quick cattle in this period.[21] But this conception of their position as integral to the household is not one that can be traced throughout England, and a study of the capital reveals that other domestic arrangements were in existence at the time Margery Reade wrote her will, domestic arrangements that did not seem to include animals in the same way. Analysis of bequeathing patterns in London reveals this different picture by showing almost nothing, but it turns out that that nothing is something worth contemplating: that the lack of evidence might, in fact, be the evidence that reveals that in this other world, the idea of kindness in its dual meaning was being transformed. And the lack of evidence can be interpreted in part because there is much that is known about the capital. Thus, the chapter begins its journey into the city with what is known and moves to speculate with the near-silence of the wills from that foundation.

By the middle of the sixteenth century, London was ten times bigger than any other city in England and its growth rate outstripped that of the rest of the country by more than two times. This population increase was not because the capital had a higher birth rate than death rate than elsewhere. Rather, "London grew by in-migration, draining the countryside of people to the tune of 6,000 a year in the late sixteenth century."[22] This represents an annual shift of about half of the population of Colchester, then the largest

---

21. A few wills in the large ERO dataset seem also to refer to the sharing of sheep, or "farming to halves." William Robinson of Little Burstead noted that "whereas Goodman Glascocke of Brome hill hath two sheepe of myne to halfes I geve him the said two sheepe, But my will and meaning is that hee shall paie the iiiis wch hee oweth to me for the wolle & lambe of the said two sheepe for the yere last past vnto my said executor"; ERO, D/ABW 44/266 (1623), made probate 140 days after writing. Mary Stacie, a widow of Springfield, bequeathed her son George "one ewe wch my sonne in law now keepeth to halves, but my pt in the lambe I give to marie his daughter"; ERO, D/ABW 50/49 (1630), made probate 76 days after writing. On this practice, see Elizabeth Griffiths and Mark Overton, *Farming to Halves: The Hidden History of Sharefarming in England from Medieval to Modern Times* (Basingstoke: Palgrave, 2009).

22. Robert O. Bucholz and Joseph P. Ward, *London: A Social and Cultural History, 1550–1750* (Cambridge: Cambridge University Press, 2012), pp. 7–8. They note that "out of every 1,000 Londoners, 35 would be born each year, but 40 would die" (p. 64). In the early modern period there was a much greater separation between the City of London and the city of Westminster than there is today. Now perceived as one enormous conurbation, it was in the early seventeenth century that The Strand was developed to form a link between the two. See Emrys Jones, "London in the Seventeenth Century: An Ecological Approach," *The London Journal* 6, no. 2 (1980): 130. In what follows, "the city" refers to the City of London.

town in Essex.[23] Obviously, not all migrants came from Essex, a county whose southwestern border with Middlesex was within about five miles of and the furthest reaches of which was less than seventy miles from London, but many undoubtedly did.[24] What drew people to the capital was the opportunities it offered. Wages for craftsmen and laborers were up to 50 percent higher in London than in the rest of the south of England, while food prices remained at or sometimes below those of the rest of the country.[25] The city also offered possibilities for training in the form of apprenticeships to trade guilds and companies. Indeed, such was the pull of the capital that it has been estimated that perhaps one in six early modern English people who survived to adulthood lived in London at some point in their lives.[26]

The growth of London and the presence there of so many new migrants has been read as inevitably undermining traditional conceptions of neighborliness upon which so much of the perceived stability of country life relied.[27] As parish records show, Margery Reade's extended family lived close to her, and even when her daughter Jane and her family moved away from West Bergholt, it was just to the village of Fordham, two miles northeast.[28] In the city, in contrast, face-to-face social relations might be transient and thus the kind of informal agreements that underpinned rural community life—such as those

23. Figure from A. P. Baggs, Beryl Board, Philip Crummy, Claude Dove, Shirley Durgan, N. R. Goose, R. B. Pugh, Pamela Studd and C. C. Thornton, "Tudor and Stuart Colchester: Introduction," in *A History of the County of Essex*, vol. 9, *The Borough of Colchester*, edited by Janet Cooper and C. R. Elrington (London: Victoria County History, 1994), pp. 67–76, http://www.british-history.ac.uk/vch/essex/vol9/pp67-76, accessed 14 January 2016. E. Anthony Wrigley estimated that the population of Colchester was ca. 5,000 in 1600. See Wrigley, "Urban Growth and Agricultural Change: England and the Continent in the Early Modern Period," *The Journal of Interdisciplinary History* 15, no. 4 (1985): 686.

24. See P. Griffiths, J. Landers, M. Pelling, and R. Tyson, "Population and Disease, Estrangement and Belonging 1540–1700," in *The Cambridge Urban History of Britain*, vol. 2, *1540–1840*, edited by Peter Clark (Cambridge: Cambridge University Press, 2000), p. 199.

25. John Chartres, "Food Consumption and Internal Trade," in *London 1500–1700: The Making of the Metropolis*, edited by A. L. Beier and Roger Finlay (London: Longman, 1986), pp. 171–72; Jeremy Boulton, "Wage Labour in Seventeenth-Century London," *Economic History Review* 49, no. 2 (1996): 287.

26. Figure from E. A. Wrigley, "A Simple Model of London's Importance in Changing English Society and Economy 1650–1750," *Past and Present* 57 (July 1967): 49.

27. See Bucholz and Ward, *London*, p. 72. See also Beier, "Social Problems in Elizabethan London," *Journal of Interdisciplinary History* 9 no.2 (1978): 221. See also Griffiths, Landers, Pelling, and Tyson, "Population and Disease," p. 232; and Vanessa Harding, "Recent Perspectives on Early Modern London," *The Historical Journal* 47, no. 2 (2004): 435.

28. In addition to their daughter Alse, who was baptized in West Bergholt in 1622, Jeremy and Jane Mills are recorded as having two other children in Fordham: John, baptized and buried in February 1625, and Grace, baptized in 1626. Parish Register of All Saints, Fordham, ERO, D/P 372/1/1 (1563–1730).

relating to the lending of a cow—might be more difficult to sustain.[29] But the shift to urban living might also have been experienced as liberating, of course, offering individuals an opportunity to escape from their likely role in their family homes where, as the bequests in Thomas Reade's will show, a son often followed his father onto his land and into his occupation and a daughter might well marry and do as her mother had done in the same village. Given this, it is unsurprising that even at the time, migrating to London was regarded as a double-edged sword. In 1632, for example, Donald Lupton wrote that the City was "the Country-mans Laborinth, he can find many things in it, but many times looseth himselfe."[30] The idea of losing oneself might be literal, of course: a countryman might get lost in the "Laborinth" of streets in the unfamiliar urban sprawl. He might also (and this was a common trope of the period) lose his wealth to the consumer opportunities that were available there. But perhaps more dangerous still is Lupton's more poetic sense that a countryman might lose that which gave him his identity—his "kindness," you could say, might be erased.

Lena Cowen Orlin has used a maritime rather than classical metaphor, writing of "unmoored patterns of social connection" in early modern London, and the image usefully conveys the potential for people to find themselves adrift from all that gave them meaning.[31] However, there were some ways the new arrival might gain a sense of anchorage in the community they had left behind: inns provided meeting points for incomers from particular areas and new migrants could call distant family networks into action too.[32] Jeremy Boulton has noted that "it was common for London kin to act as hosts for immigrant spinsters,"[33] and London wood turner Nehemiah Wallington's writings reveal that it was not just women who used such ties. He recorded how, in the 1640s, his Irish-born nephew Charles, as well as his widowed sister-in-law Sarah and her two children from Lincolnshire, had come to live with

---

29. It is perhaps in this context that the literacy rates in London are worth noting. In the large ERO dataset 25 percent of wills are signed; in London, 45 percent of wills are signed.

30. Donald Lupton, *London and the countrey carbonadoed and quartred into seuerall characters* (London: Nicholas Okes, 1632), p. 2.

31. Lena Cowen Orlin, "Temporary Lives in London Lodgings," *Huntington Library Quarterly* 71, no. 1 (2008): 236. Orlin is writing here about the precarious nature of women's lives in the capital, but I suggest that her point also holds true for men.

32. J. A. Chartres, "The Capital's Provincial Eyes: London's Inns in the Early Eighteenth Century," *The London Journal* 3, no. 1 (1977): 31–32.

33. Jeremy Boulton, "London 1540–1700," in *The Cambridge Urban History of Britain*, vol. 2, *1540–1840*, edited by Peter Clark (Cambridge: Cambridge University Press, 2000), p. 345.

him, his wife, and their daughter in their Eastcheap home "with its two bed-chambers and garret." Charles became Wallington's apprentice.[34]

But, as well as underlining the family ties that could persist in the capital, the crowded living conditions of Wallington's household give another sense of a difference between London and the surrounding countryside. The city grew by 40,000 people in the 1620s (despite the devastating outbreak of the plague in 1625 that killed up to 40,000), and in the next decade, the population increased by 56,000. This rapid growth put inevitable pressure on housing.[35] In rural England, Elizabethan law required that a cottage (the smallest holding, often inhabited by laborers) be accompanied by a minimum of four acres of land, a holding that was believed to be enough to sustain a household.[36] This did not always happen, and some dwellings had much less, but the expectation is telling. Thomas Reade the younger's family, for example, grew up on a holding of well over 100 acres; and of the twenty-six other West Bergholt residents listed in the 1621 survey, only six possessed less than four acres—and the survey was just of the manor owned by Sir John Denham, so it did not include all the land in the village.[37] London changed things: the density of houses in "the border parishes" of the capital was fifteen per acre, but in the city there could be as many as ninety-five.[38] Indeed, some migrant families found themselves living not only in close proximity to their neighbors, but in sheds—that is, lean-to wooden structures, sometimes only twelve feet square—or in divided houses.[39] In 1637, for example, a government inquiry found "10 several families, divers of which also had lodgers" living in one ten-room house in Silver Street, just inside the northern boundary of the city.[40]

34. Paul S. Seaver, *Wallington's World: A Puritan Artisan in Seventeenth-Century London* (London: Methuen, 1985), pp. 82, 84.

35. William Baer, "Housing for the Lesser Sort in Stuart London: Findings from Certificates and Returns of Divided Houses," *The London Journal* 33, no. 1 (2008): 63. The wills in the LMA dataset record this devastating plague, not only in the higher number of wills made during 1625 (241, in contrast to an annual average of 68 for the other years) but also in the higher proportion of nuncupative wills made that year. In all years but 1625, nuncupative wills made up 18.5 percent of London documents; they made up 24.5 percent of London documents in 1625.

36. See figures in Craig Muldrew, *Food, Energy, and the Creation of Industriousness: Work and Material Culture in Agrarian England, 1550–1780* (Cambridge: Cambridge University Press, 2011), pp. 110–11.

37. ERO, D/DMa M18. "West Bergholt: Manors and Other Estates" lists another manor in the village alongside Denham's land. In *A History of the County of Essex*, vol. 10, *Lexden Hundred (Part) Including Dedham, Earls Colne and Wivenhoe* (London: Victoria County History, 2001), http://www.british-history.ac.uk/vch/essex/vol10/pp2730, accessed 25 September 2017.

38. Figures from Laura Gowing, *Domestic Dangers: Women, Words, and Sex in Early Modern London* (Oxford: Oxford University Press, 1996), p. 21.

39. Baer, "Housing for the Lesser Sort in Stuart London," p. 68.

40. Cited in Vanessa Harding, "Space, Property and Propriety in Urban England," *Journal of Interdisciplinary History* 32, no. 4 (2002): 567.

This lack of living space might have meant that separation from a familiar community was juxtaposed with overcrowding amid strangers, and so the rapid population expansion had potentially less tangible connotations, with an impact beyond the space individuals and families found themselves occupying: the concepts of private and public were also changed. Laura Gowing has noted that in depositions given to the Consistory Court—the court that dealt with moral crimes—London residents referred to the poor quality of buildings, to holes in walls or to thin walls.[41] These were sometimes cited as exposing domestic life to neighborly surveillance with the seclusion of the home undermined by the scrutiny and eavesdropping of unfamiliar—perhaps better termed unkind—people. London changed family life.

Another outcome of the massive population increase in the city brings us to think about the notion of kindness in another way, because in the capital the line between human and animal also shifted. There was an established joke in the period that country folk were closer to animals than their urban(e) neighbors, but the reality for many Londoners was rather different. Vanessa Harding has found that in the 1630s in Aldgate (the entry point into the city for those coming from Chelmsford, according to John Taylor's 1637 survey of carriers), people were living in converted stables; and cow-houses and pigsties were also being adapted to human habitation.[42] Thus, where the accommodation of sons and male servants was often above animal stalls in the countryside, in London the housing situation forced people into the place of animals.[43] In addition, as the population grew and human accommodation encroached upon the open areas in the city, fewer homes had outside space: a parliamentary survey from 1651 found that only 33 percent of properties in Tower Liberty and Shadwell, just east of London, had gardens.[44] And there is evidence that food production

---

41. Gowing, *Domestic Dangers*, pp. 56, 71; see also Harding, "Space, Property and Propriety in Urban England," p. 566. Citing Matthew Johnson's research, Jane Whittle notes that the "closed house" with its "specialization of activities within particular rooms" began to emerge from 1560 and led to a "segregation of household members." Housing in early seventeenth-century London appears to undo such separation. Whittle, "The House as a Place of Work in Early Modern Rural England," *Home Cultures* 8, no. 2 (2011): 136.

42. John Taylor, *The Carriers Cosmographie* (London: A. G., 1637), n.p.; Vanessa Harding, "Families and Housing in Seventeenth-Century London," *Parergon* 24, no. 2 (2007): 133; Baer, "Housing for the Lesser Sort in Stuart London," p. 63. Such conversions were not just happening without the knowledge of the city authorities, it would seem. The Chamber of the City of London granted Elizabeth Knight, a widow, "a lease for xxi yeares of one tent [tenement] and other roomes sometymes vsed for stables" in July 1589. "City Land Grant Book," LMA, CLA/008/EM/02/01/001 (1589–1616).

43. See Edwin J. Rose, "Man Set over Animals," *Vernacular Architecture* 29, no. 1 (1998): 21; and M. W. Barley, *The English Farmhouse and Cottage* (London: Routledge and Kegan Paul, 1961), p. 11.

44. M. J. Power, "The East and West in Early Modern London," in *Wealth and Power in Tudor England: Essays Presented to S. T. Bindoff,* edited by E. W. Ives, R. J. Knecht, and J. J. Scarisbrick (London: University of London, Athlone Press, 1978), p. 170. Power notes that inhabitants of Tower Liberty

that might have taken place outdoors if room had been available was moved inside into the domestic space: rabbits were being intensively bred for meat and fur in London lofts, cellars, and outhouses.[45] So, where Markham's husbandman left his quick cattle in the "beast house" in the evening, a Londoner might find his or her own home overrun by animals.[46]

But it was not just the architectural aspects of the rapid growth in London's population that shifted the boundaries between humans and animals in the city; life in the early modern capital was distinct from elsewhere because the environment was one in which encounters with quick cattle were also very different. Here we do not find the day-to-day contact that was so central to life for many in the countryside: instead, for the majority of Londoners, work was done in what we might call species isolation and food stuffs were purchased rather than produced through labor alongside livestock.[47] This difference is reflected in the wills left by inhabitants of the Diocese of London in the period 1620–1634.

The complete, clean set of wills from the Diocese of London for this fifteen-year period lacks any documents from 1623—none have survived—and totals 2,530. It includes 517 wills by individuals who were not residents of London or of parishes in Middlesex or who gave no place of residence and, in order to focus on London and Middlesex, therefore, the dataset that will be the basis of the following discussions excludes these documents. Those omitted are: 179 wills from towns and villages in the area of Essex that are under the authority of the Diocese of London, twenty-eight wills from testators who listed their place of residence as elsewhere in England, six wills from residents of Scotland, six wills from overseas,[48] and 133 wills that do not

---

"could have obtained some relief by escaping to the large open spaces of Little and Great Tower Hill" (p. 171).

45. Malcolm Thick, "Intensive Rabbit Production in London and Nearby Counties in the Sixteenth, Seventeenth, and Eighteenth Centuries: An Alternative to Alternative Agriculture?" *Agricultural History Review* 64, no. 1 (2016): 12. Thick records that Sir Hugh Plat noted in the 1580s "that two does could produce 33 offspring a year whose skins could be sold for 2s. or 3s. each" (p. 3). In 1562, Italian visitor Alessandro Magno noted that there were rabbits "in abundance" in the capital. Caroline Barron, Christopher Coleman, and Claire Gobbi, "The London Journal of Alessandro Magno 1562," *The London Journal* 9, no. 2 (1983): 143.

46. Gervase Markham, *Markhams Farewell to Hvsbandry* (London: I. B. for Roger Jackson, 1620), p. 145.

47. John Styles notes that "even [the] poor depended on commercial mechanisms for the supply of basic essentials like food or linen to an extent that was unparalleled in rural areas." Styles, "Product Innovation in Early Modern London," *Past and Present* 168 (August 2000): 129. The notion of "species isolation" refers only to working practices: people continued to share their homes with rats and fleas (at least), as the plague outbreaks evidence.

48. Two are from the Netherlands, one each from Ireland, Denmark, Japan, and Virginia Plantation. The testator who listed Japan as his place of residence was a mariner. The LMA dataset includes

specify the testator's place of residence (92 of which are by sailors, mariners, and those who listed themselves as going on a voyage).[49] I also omitted 165 wills by seabound testators who listed their place of residence as a particular ship because their permanent place of residence is not known. Apart from the Diocese of London wills from Essex parishes, which resemble those in the large ERO dataset in that 12 percent include specific animal bequests,[50] only one of the other documents excluded contained a specific animal bequest—the 1622 will of John Jane, a sailor who gave his dwelling as the ship *The Charles*. He bequeathed to his shipmates John Betchered and Nicholas Johnson "one Civet cat equally between them."[51] The animal would have originated in South Asia, and civet for use in perfume and medicine could be "harvested" from a sac near to its anus. Whether Jane's animal survived its voyage is not known, but some did. In 1604 an ounce of civet was worth £1 6s 8d in England.[52]

The exclusion of these wills from the LMA dataset leaves 2,013 documents. Of these, 1,130 (56 percent) are by testators living in City of London parishes and the remaining 883 (44 percent) are by testators from the surrounding suburbs of the capital in Middlesex—such as Whitechapel in the east, Islington in the north, and St Giles in the Fields in the west and from what were then more rural areas of the county such as Tottenham, Ruislip, Edmonton and Brentford (the diocese does not include any parishes south of the Thames). The chapter will look at these wills later, but its first focus is on those by testators who listed their place of residence as the City of London. The wills, like the architecture, give a picture of the place of animals in the capital that is very different from what is found in Essex.

In the large ERO dataset, just over 10 percent of wills included specific animal bequests. In the wills by residents of London that figure is very different: less than 1 percent. That is, only 11 of 1,130 wills made probate in the Diocese of London between 1 January 1620 and 31 December 1634 that were written by testators who listed their place of residence as one of the city's parishes or named themselves citizens of London included any specific animal bequests.

wills by foreigners who listed themselves as residents of London: nine are by testators who originated in the Netherlands and ten are by individuals from France.

49. When the testator did not list a place of residence but requested burial in a particular church, I have taken that to be the place of residence.

50. I return to the issue of why this figure is slightly higher than in the rest of Essex later in this chapter.

51. LMA, DL/C/B/008/MS09172/033 119 (1622), made probate 196 days after writing.

52. See Karl H. Dannenfeldt, "Europe Discovers Civet Cats and Civet," *Journal of the History of Biology* 18:3 (1985), pp. 403–31. The value of civet in England is given on p.411.

Many other testators used the collecting phrases we are familiar with, of course, and one—Thomas Derds, a barber surgeon—referred to his "quicke Cattle," but the lack of specific reference to animals marks the London wills as different from those from rural Essex.[53] On its own this does not automatically mean that there were proportionately fewer people in possession of animals in the capital than in Essex, of course (many wills, as we have seen, did not include all the testator's possessions), but other details about the nature of urban life that are visible in the wills offer further evidence to support that suggestion.

According to the occupational status people gave themselves or were given by their neighbors, in Essex over 43 percent of testators were involved in agricultural production as their primary occupation.[54] In London, that proportion was just over 1.5 percent. That is, whereas in Essex those terming themselves yeomen and husbandmen made up, respectively, 25 percent and 17 percent of testators, with laborers making up just over 1 percent, in the capital only eighteen testators termed themselves yeomen and there were no testators who termed themselves husbandmen or laborers. The reason for this is obvious: many people who lived in the city did not have access to pastures or outside space, and so their livelihoods were made in different ways, with these occupations reflecting the distinct environment of the capital. For example, twenty-nine London testators listed themselves as haberdashers, whereas in the large ERO dataset there were only three. In the capital, three times more testators were involved in the sale of food and drink—innkeepers, brewers, cooks, vinters—than in the whole of Essex and five times more were butchers, poulterers, fishmongers, fruiterers, grocers, and bakers. Also included in the London dataset are wills by merchant tailors, stationers, goldsmiths, embroiderers, plasterers, pewterers, girdlers (beltmakers), armorers, plumbers, a clockmaker, a coppersmith, a hatband maker, a printer, and a sugar baker. None of these professions appear in the large ERO dataset. This worked the other way around, too: occupations other than husbandman and laborer appear in Essex but not London. These include millers, wheelwrights, maltsters, plowwrights, potters, palemakers (fence makers), shearmen, shepherds, and one skepmaker.

As well as the different range of occupations in the capital, it is also notable that while just under 20 percent of wills in the large ERO dataset were by women, over 32 percent of wills in London were by female testators, with

---

53. LMA, DL/C/B/008/MS09172/036 22 (1625), made probate 28 days after writing.

54. It is likely, of course, that a high proportion of other testators were involved in agriculture as a second occupation, but that is difficult to assess.

widows making up just under 25 percent of the London wills overall; the figure was just under 17 percent in Essex. This may reflect the fact that women were simply more likely to write wills in London,[55] but it is also possible that the proportion of wills by female testators reveals that it was easier for a lone woman to run a business and a household in the city than in a rural community, perhaps because in the country the emphasis on agricultural production made the need for male labor (and thus remarriage) more pressing. As such, the high proportion of widows might be used to support the evidence of the range of occupations among the capital's male testators in noting the absence of agricultural employment in the city.

Thus, while the lack of specific animal bequests in wills cannot on its own prove a lack of livestock in the capital, when set alongside other evidence—about the transience of the city's population, about housing (with animal spaces being transformed into human abodes and with lack of access to gardens and fields for many), and about the occupational range of testators—the wills do support the assumption that fewer testators living in London were in possession of quick cattle than in Essex. (The wills do not offer any evidence for animals such as pets, of course.) Given these dissimilarities it is unsurprising that instead of offering some residual reflection of rural testaments, an analysis of the eleven London wills that do include specific animal bequests further reveals how likely it is that the residents of the capital lived very differently from their country neighbors. It is not simply the small number but the content of the documents that shows the capital to be a world bereft of close contact with quick cattle.

Six of the eleven London wills (54 percent) that include animals refer only to horses: in the large ERO dataset the proportion was very different, with 51 of 551 wills (9 percent) with specific animal bequests containing horses alone. The horse bequests in London were made by Richard Lawrence, a citizen and woodmonger who bequeathed to his wife's son Thomas "one of my carr poonies and one of my Carres and my Bald heyde horse" and to his daughter Margaret "my other carr ponie and one of my Carrs and my graye horse";[56] Thomas Thorpe, citizen and white baker (i.e., a baker of white bread) who left his brother Robert "my grey mare";[57] Christopher Phillips, of St Sepulchre without Newgate, one of the few who termed himself a

---

55. Comparative literacy statistics are noteworthy here: in Essex, just under 6 percent of wills by women were signed and less than 5 percent of all signed wills were by female testators, while in London 17 percent of wills by women were signed and over 11 percent of all signed wills were by female testators.

56. LMA, DL/C/B/008/MS09172/033 166 (1622), made probate 21 days after writing.

57. LMA, DL/C/B/008/MS09172/037 110 (1626), made probate 36 days after writing.

yeoman, displayed his wealth when he bequeathed to his kinsman Henry Blunt, "my eight coache horsses and all the harnesses and furniture to them belonginge";[58] Ralph Criple, another citizen and woodmonger, who left his son Thomas his "horse and Carr";[59] and Francis Man from St Dunstan in the West, another yeoman, who gave his cousin Richard "of great Jenkins in the parish of Hollingbury Magna [Great Hallingbury] in the . . . county of Essex my gray Nagg wch hee was to giue mee fiftie shillings for."[60] The other horse bequest is in the will of Robert Richardson, citizen and scrivener (scribe), which included property with appurtenances in Lancashire and, in a bequest to his parents, "My Mare and ffoale wch are kepte at Sandscale" in Cumbria.[61]

Testators of other of the remaining five London wills with specific animal bequests also reveal that their quick cattle were kept outside the capital. In her 1627 testament, the widow Rebecca Pendred, a resident of St Sepulchre without Newgate, noted that her late husband's will had entailed their London property to their son John. She herself bequeathed jewelry and apparel and left to "Sara Standish daughter to Hugh Standish Carpenter and to Mary Heward daughter of Thomas Heward deceased my stock of Bees at Kensworth in the county of Beds nowe in possession of Thomas Gregory."[62] The will makes no other mention of Gregory or of property in Bedfordshire. Another widow, Cicely Duncombe, who listed herself as "of London" and wished to be buried in St Stephen Coleman Street, left legacies totaling over £1,400 in her 1634 will, and bequeathed to her daughter Dame Sarah Roe "all and singular my horses and cartes wth the furniture thervnto belonging. All Kinne Wendlinges hogges sheepe and other cattle whatsoeu' as well small as greate as I the said Cicely shall haue, or be owner of at the time of my decease, either at Musselhill, or at any other place whersoeu'."[63] Muswell Hill is nearly six miles north of the city. Somewhat less grandly, thirteen years earlier, Thomas Caton, one of Rebecca Pendred's neighbors in St Sepulchre without Newgate, had left "all the residue of all & singular my goods Chattells sheepe debts & psonall estate whatsoeuer & wheresoeuer" to his wife Ellyn: the "wheresoeuer" hinting at the possibility that at least some of his possessions might be beyond the parish.[64]

---

58. LMA, DL/C/B/008/MS09172/037 251 (1626), made probate 45 days after writing.
59. LMA, DL/C/B/008/MS09172/039 70 (1629), made probate 981 days after writing.
60. LMA, DL/C/B/008/MS09172/039 192 (1630), made probate 17 days after writing.
61. LMA, DL/C/B/008/MS09172/035 62 (1625), made probate 28 days after writing.
62. LMA, DL/C/B/008/MS09172/038 73 (1627), made probate 15 days after writing.
63. LMA, DL/C/B/008/MS09172/042 59 (1634), made probate 1,080 days after writing.
64. LMA, DL/C/B/008/MS09172/032 70 (1621), made probate 48 days after writing.

Thus, of the eleven wills from London that include specific animal bequests, four refer to animals that were (or were heading) outside of the city—the bees in Kensworth, the livestock in Muswell Hill, the horse in Cumbria, and the horse that was going to and may already have been in Essex.[65] In addition, the language of another will raises the possibility that its sheep were also elsewhere—although the Agas map of London from 1561 does depict a cow grazing on a field in the north of the parish in which Caton lived, so it is not impossible that his sheep were kept close by.[66] Of the remaining six wills, the fact that Christopher Phillips described his animals as "coache horsses" means, I think, that they were unlikely to have been used for labor in the fields,[67] and it is probable that the two woodmongers and the white baker who listed horses and ponies among their property used them for their trade or for transport rather than agriculture. This means that it is plausible that only two of the 1,130 wills from London in the fifteen-year period the LMA dataset covers include specific references to animals that were engaged in something like agricultural production in the city. This is scant evidence on which to build any narrative of life with quick cattle in the early modern capital, but analysis of these two wills opens up ways of thinking about the presence and absence of livestock in England's largest city. Questions of kindness—of care and of identification—remain central.

The two London wills I have not yet discussed that include specific animal bequests both refer to cows. The earliest is from 1622 and is that of the widow Anne Harrison of St Botolph without Bishopsgate in the east of the city. Her will included only three legatees: the "gardner" Walter Embers, "my servant Amy" (a gap is left where the scribe has left out but not returned to fill in Amy's family name), and "my cussen Mary Sheres now allso my servant." In addition, Harrison left the poor of the parish "ffortie dozen of bread to be distributed at my funerall" and she named as her executor the

---

65. To complete the picture: one Essex testator bequeathed a horse to a legatee who was described as living in London. Richard Freebody, a gentleman from Little Stambridge, left Eleanor "the wife of my brother John Ffreebody of London one browne amblinge nagge of three yeares old"; ERO, D/ABW 43/71 (1621), made probate 25 days after writing. Andrew Reynolds, a Romford butcher, bequeathed his father "my graye horse and my Stall att London," but it is possible that it was only his stall (shop) that was in the city and that the horse was in Essex; ERO, D/AEW 17/139 (1623), made probate 27 days after writing.

66. The Agas map (a map previously ascribed to the cartographer Ralph Agas) can be found on https://mapoflondon.uvic.ca/agas.htm. No animals are visible in the fields in John Norden's 1593 map, but the fields remain: see https://commons.wikimedia.org/wiki/File:London_-_John_Norden%27s_map_of_1593.jpg. Both accessed 29 September 2016.

67. Coaches first appeared in London in 1564. See Bucholz and Ward, *London*, p. 40.

baker John Weekes, a "loving freind" from East Smithfield, perhaps the man who would supply the bread for the poor. In the will's specific bequests, Embers was forgiven his debt to her of eighteen shillings; Amy was left "my biggest Iron pott, my greate ketle and one long black chist"; and Mary "my two litle potts and one wainescott chist." In addition, Harrison stated that "I doe give and bequeath vnto the said Amy and Marie all such debts as are to me due or shalbe due vnto me for milke."[68] This is the only bequest in either the LMA dataset or the large ERO dataset that includes any reference to milk. Other wills mention milkhouses (i.e., dairies), dairy and cheese-making equipment (as in Margery Reade's bequest from her husband),[69] and references to cheese can also be found (as in Christopher Fuller's will); but milk itself—a quickly spoiled substance—is neither bequeathed nor referred to elsewhere. Having noted the money she was owed for milk, Harrison then requested that if, having paid off her own debts and covered the costs of her funeral, her executor finds "that here shalbe left of my estate, soe much as shall amount vnto as greate a share for my executor as I shall hereby give vnto the said Amy & Marie, Then I do give vnto each of them one Cowe over & besides the legacies wch before I have herein given." The (somewhat opaque) implication is that Harrison has at least two or three, and perhaps more, cows.[70]

A nostalgic moment in John Stow's *Survay of London* of 1598 offers a possible context for understanding widow Harrison's bequests. He recalls a "Farme" close to the Minories, a former abbey that may have been less than half a mile from Harrison's residence in St Botolph without Bishopsgate, "At the which Farme," he wrote,

> I my selfe in my youth haue fetched many a halfe pennie worth of milke, and neuer had lesse then three Ale pints for a half penie in ye sommer, nor lesse then one Ale quart [two pints] for a halfpennie in the

---

68. LMA, DL/C/B/008/MS09172/033 38 (1622), made probate 36 days after writing.

69. Among the other detailed descriptions is that of the Bradwell-juxta-Mare widow Mary Garrett, who bequeathed to her son John "two Cheese Mootes [vats] wch may be most ffittingest for him, one Cheese bread [a board used to press out the whey], three Milke keelers [vats] one brasse milke pann with out ~~the~~ ^a^ hoope, one little longe Cherne ffurnisht one Tankerd Cherne, two ^milke^ pailes." In addition, he received "two loade of Cheese one ffirkin of Butter" and "some ffoure of my Indifferent Cowe." "Indifferent" here means "any among them" rather than mediocre. ERO, D/AEW 17/156 (1623), made probate 30 days after writing.

70. LMA, DL/C/B/008/MS09172/033 38. There is a lack of clarity as to whether the statement means that John Weekes should receive "as great a share" as each of the servants individually or that he should receive as much as what Amy and Marie got between them. Harrison may, of course, also have had more than three cows: the will does not make clear what she was owed or what possessions (if any) might need to be sold to clear her own debts before the executor could take his share.

winter, alwaies hot from the cow, as the same was milked and strained. One *Trolop*, and afterwardes *Goodman*, were the Farmers there, and had 30. or 40. kine to the payle.[71]

Stow would have walked less than a mile from Cornhill, where he was brought up, to get this still-hot milk (its being hot evidence that it had not been adulterated, perhaps[72]), and it makes sense that, unlike cheese, it had to come from close to or within the city; otherwise it would go off before sale. It is in this urban context that widow Harrison might have run a small dairy, and her proximity to Spitalfields, an area of open ground—again pictured with cows on it in the Agas map—might mean that she had access to pasture for her animals.[73] Indeed, by the 1620s, Harrison might have taken over some of the business of the dairy that Stow recalled, as, at the end of his reminiscing, he noted that the son of Goodman, the owner of the farm at the Minories, "being heyre thereof, let out the ground first for grazing of horse, and then for garden plottes, and liued like a gentleman thereby."[74] The garden plots did not last, however, and in his update of Stow's work from the early eighteenth century, John Strype revealed their fate: "now *Goodman's Fields* are no longer Fields and Gardens, but Buildings consisting of many fair Streets . . . and a large Passage for Carts and Horses out of *Whitechapel*."[75] By 1720 the "Farme" had become a highway.

It is possible, then, to return to the 1620s, that Anne Harrison might have had a business selling milk to London residents. The reference in her will to the "debts as are to me due or shalbe due vnto me for milke" could refer to financial arrangements she had come to with customers who had bought from her on credit (among whom, perhaps, was Walter Embers the gardener), and she may have engaged her two servants Amy and Mary to do the milking alongside other household chores. Harrison's small urban herd of a few cows might thus have provided an income to her as well as much-needed fresh milk to some of her neighbors. Alternatively, if she had a larger herd than those

---

71. John Stow, *A Svrvay of London* (London: Iohn Wolfe, 1598), p. 91.

72. See Hannah Velten, *Beastly London: A History of Animals in the City* (London: Reaktion Books, 2013), p. 30. The nineteenth-century practices Velten wrote about may have been common two centuries earlier.

73. There is no reference to real estate (as opposed to personal estate) in Harrison's will, and her collecting phrase includes reference only to "the rest & residue of all my goods & chattells." LMA, DL/C/B/008/MS09172/033 38.

74. Stow, *A Svrvay of London*, p. 91. "Mr William Goodman Gentleman of the Mynories Street" buried a daughter in St Botolph, Aldgate, on 28 February 1621. Parish Register of St Botolph, Aldgate, accessed via ancestry.co.uk on 21 October 2016.

75. John Strype, *A survey of the cities of London and Westminster* (London: A. Churchill, 1720), Book II, p. 15.

cows specifically referred to in her will (if her will can be called specific on that matter), the money she was owed for milk could reveal contracts she might have had with independent milkmaids. Such milkmaids would have been self-employed women or women employed by a distributor who undertook with Harrison to buy a certain amount of milk. These maids would milk the cows and then sell the milk on the capital's streets, keeping the profits for themselves or their employer.[76] Such an arrangement would guarantee that Harrison would sell her cows' milk immediately and thus waste little. It would also ensure that the milkmaids would have a constant supply of fresh milk to sell "hot from the cow," as Stow put it.

Large dairies were to be found in the countryside, of course—the reference to "seauenscore" (140) cows being milked by his mother and her maids by Bartholomew Dowe's Suffolk man reveals this.[77] But in London, milkmaids might not live within the household and do the milking alongside other domestic tasks; they could have come in from outside and have used the sale of the milk they took from the cows as their sole means of earning a living.[78] Such an arrangement seems to be visible in the 1627 will of Phillipp Pitt, a yeoman of Stoke Newington, three miles north of London, for example. His will, registered under the jurisdiction of the Prerogative Court of Canterbury rather than the Diocese of London,[79] included a bequest "vnto all my Milkers Tenne shillinges a peece that are Milkers at my departure." Those he termed his "servants" were differentiated from the "milkers" and received bequests

76. See P. J. Atkins, "London's Intraurban Milk Supply, circa 1790–1914," *Transactions of the Institute of British Geographers* 2, no. 3 (1977): 392; P. J. Atkins, "The Retail Milk Trade in London, c. 1790–1914," *Economic History Review*, new series 33, no. 4 (1980): 523; and David Pennington, "Taking to the Streets: Hucksters and Huckstering in Early Modern Southampton, circa 1550–1652," *The Sixteenth Century Journal* 39, no. 3 (2008): 664.

77. Bartholomew Dowe, *A dairie Booke for good huswiues* (London: Thomas Hacket, 1588), A3r. Most cheese came from Suffolk and Essex; but milk was also brought to the City from Middlesex. See F. J. Fisher, "The Development of the London Food Market, 1540–1640," *The Economic History Review* 5, no. 2 (1935): 51, 55.

78. On the work of milkmaids, see Jane Whittle, "Housewives and Servants in Rural England, 1440–1650: Evidence of Women's Work from Probate Documents," *Transactions of the Royal Historical Society* 6th series 15 (2005): 62.

79. Wills under the jurisdiction of the Prerogative Court of Canterbury (PCC) include property in two or more locations and, frequently, wealth that was uncommon in wills in the large ERO and LMA datasets. In a sample of 161 of these wills from Essex, Middlesex, and London dating from 1627, the middle year of the other datasets I have used here, knights, esquires, and gentlemen make up over 20 percent of testators in the PCC, while in the large ERO dataset (where there is also a baronet) they make up just over 2.5 percent and in the LMA dataset just over 3 percent. Only eleven PCC wills (7 percent) include specific animal bequests, and five of those eleven (45 percent) are bequests of horses that may not reflect agricultural production. If that was the case, then only six of 161 PCC wills—4 percent—include specific bequests of agricultural animals. This in a sample in which 51 percent of testators referred to real estate (the equivalent figure in the large ERO dataset is 45 percent).

of 20 shillings each.[80] Pitt did not specify any animals in his will, but the distinction between milkers and servants hints at the existence of a commercial dairy enterprise and the different size of the bequests to those in the two groups may signal that the milkers were regarded as less valued members of the household-family than servants. They were, perhaps, more likely to be employees with whom the master would have had less contact.

Such commercialization of dairying in London not only impacted the engagement of the employer with his workers, though; it would also have led to changes in the maid's relationship with the cows. First, much of the maid's day would have been spent on the streets far away from the dairy, and just as her contact with her master might have been less than what it would have been in a rural household, where she would also have lived within the household-family, so her contact with the cows would also have decreased: there would have been less opportunity for the passing acknowledgments, greetings, pats that might have been part of the working day on a smallholding, for example.[81] Displays of kindness, in short, would have declined. Second, commercialization would mean that the milkmaid who contracted to take a certain amount of milk did not need to be concerned with the long-term welfare of the cows she milked, because for her one cow could simply be replaced by another and her contract still be fulfilled. This could mean, consequently, that such milkmaids might be highly skilled—capable of negotiating with any cow—but that they perhaps had to do so with a different sense of those creatures, because they would not necessarily have the opportunity to engage in long-term conversations with specific animals. The care they would practice would thus be more universalizing.[82] That is, all cows might be expected to engage in the partnership (if it was still viewed as a partnership) in the same way, and their individuality would be of little value—indeed, it might be viewed as a problem. The commercialization of the milkmaid, as such, can be seen as the first step on the way to the mechanization of milking, with cows being reduced to the status of fleshy machines for milk production—being, in short, un-kinded. How milkmaids who might have learned their

---

80. NA, PROB 11/151/452 (1627), made probate 8 days after writing.

81. This point should not be exaggerated because if cows in Essex were kept on a common, they would have been away from the maids for a great deal of the day.

82. Annemarie Mol and her colleagues differentiate medical ethics from care ethics in a way that is useful here. In the former care is premised on universal principles that pre-exist actual encounters between parties and medical ethics are thus the always-applied rules that engagements operate through. In care ethics, in contrast, care is localized and tentative; it reflects "attuned attentiveness and adaptive tinkering." Annemarie Mol, Ingunn Moser, and Jeannette Pols, "Care: Putting Practice into Theory," in *Care in Practice: On Tinkering in Clinics, Homes and Farms*, edited by Annemarie Mol, Ingunn Moser, and Jeannette Pols (Bielefeld: Transcript, 2010), pp. 13–15.

trade on rural farms or smallholdings before moving to the capital might have felt about the change is not recorded, but their understanding of milking as an exercise in collaboration—as an act of recognition and generosity on the part of both parties—might, in these new circumstances, have been more of a problem than a boon: milking "neatly" would need to become an act of efficiency (tidiness) rather than empathy (becoming calf).[83] That ambiguous bequest of cows in Anne Harrison's will, therefore, affords a possible glimpse of a new way of being with animals in which kindness—a sense of shared nature—might have been declining, not because people were unkind but because circumstances changed to make enacting kindness more difficult.

Where Harrison's will lacked clarity in its description of how many cows she owned, there is another kind of ambiguity in the last of the eleven London wills that includes a specific animal bequest. This is found in the 1624 will of John Rile in St Dunstan in the West, a parish on the other side of London from Harrison's residence in St Botolph without Bishopsgate. In this document Rile bequeathed "vnto my . . . sister Ellen that cowe wch is due vnto me by my said father as by his specialitie appeareth."[84] A "specialitie" in this context is a bond or promise, and the fact that it "appeareth" would hint at the possibility that the promise was written down somewhere rather than made by word of mouth. But, whether the "specialitie" was written or verbal, it is clear that the cow in John Rile's will was not an animal that he had in his possession in the city but was one that was his by right. There is no information about Rile's occupation in his will, although the parish register of St Dunstan in the West records a John Ryley who was an "instrument maker" as having had children baptized in January 1606 and April 1608.[85] If Rile and Ryley are the same person (and early modern spelling would allow for this—indeed, Rile signs his will "John Ryle"), then the fact that there is no record of his baptism in the parish register raises the possibility that he was an immigrant to the city. It also makes possible that he was not engaged in agricultural production as his primary occupation. In addition, Rile's will does not refer to any residence except the

---

83. Bernard Rollin discusses how individuals from family farms in modern America struggle to deal with the lack of personal engagement when they move to work in industrial agricultural holdings. Rollin, "The Ethics of Agriculture: The End of True Husbandry," in *The Future of Animal Farming: Renewing the Ancient Contract*, edited by Marian Stamp Dawkins and Roland Bonney (Oxford: Blackwell, 2008), p. 8.

84. LMA, DL/C/B/008/MS09172/034 25 (1624), made probate 348 days after writing.

85. LMA, P69/DUN2/A/002/MS10343, Parish Register, St Dunstan in the West (1558–1622), accessed via ancestry.co.uk, 21 October 2016. In addition to Ann (baptized 1606, buried 1616) and Thomas (baptized 1608), John the son of John Riley was baptized in the parish on 9 August 1617 and buried on 18 February 1618. The father of this latter child was not named as an instrument maker in the register.

one in St Dunstan in the West and does not state where his sister or his father lived, so it is difficult to know whether his cow was ever likely to come to the city. She may have been kept outside London by his father and Rile may have had no intention of ever bringing her to live with him; Rile—like widow Duncombe and her livestock in Muswell Hill—may have owned the animal from a distance. If this was the case, then he would be unlikely to consider the cow as an individual; it was simply one of his possessions, like the "goods & ymplements . . . bedding and whatsoeuer ells" he also bequeathed to his sister.[86]

However, there is another possibility: this animal could have been due to come to Rile to supply him with milk for sale or personal use had his death not intervened—and there are gardens visible behind the parish church of St Dunstan in the West in both the Agas and later Norden maps in which a cow might, perhaps, be kept.[87] If not, maybe the practices Eric Kerridge recorded from a later period might have existed in the early seventeenth century. He noted that in the eighteenth century, some London-based cow keepers kept their animals "confined throughout the year in dark, close sheds and stalls that promoted pulmonary consumption."[88] Little has been written about the keeping of cows within the capital in the seventeenth century,[89] but one 1612 record from the Burgess Court of Westminster cited by P. J. Atkins seems to reveal something similar to what Kerridge reported happening nearby and to show another change in the use of space. The court records one Thomas Farmer keeping "eight kyne tyed up in the kitchen which by his own confession he hath lately converted to a cowhouse."[90] Kitchens at this time were often outside the main house and Farmer perhaps had, like the shed builders of east London, done what he could to swiftly grab a new commercial opportunity that the rapid growth of the city had created. As widow Harrison knew, selling fresh milk in London was good business.[91]

---

86. LMA, DL/C/B/008/MS09172/034 25.

87. Records of the transfer of property from 1614 to citizen and grocer William Drewe note the sale of "three messuags or tenemts wth thapp'tenncs" in St Dunstan in the West, so it is possible that such properties had grazing rights, the legal meaning of appurtenances. "Bonds of Indemnication and Releases," LMA, P69/DUN2/B/061/MS03795 (1614). It is noteworthy that the testators of the three wills that might refer to livestock living in the city—Caton, Harrison, and Rile—all lived in parishes that abutted or included open space.

88. Eric Kerridge, *The Agricultural Revolution* (1967; repr., Abingdon and New York: Routledge, 2013), p. 178

89. When they attend to such animals, histories of the city focus on the fact that they were brought in from the rest of the country for sale and slaughter. See, for example, Velten, *Beastly London*, pp. 30–33.

90. Quoted in Atkins, "London's Intraurban Milk Supply," p. 395.

91. A dispute over tithes of the Black Mule, a property in St Dunstan in the West, from 1624 reveals the existence of a kitchen in "the back pte of the said tenemt" in the late sixteenth century. In

So, just as the lives of the shed-dwelling humans in the capital were very different from those in the countryside, so were the lives of the shed-dwelling cows. In rural areas, a cow's productive life could be fifteen years long, whereas in London, Kerridge notes, the cows were "fed on stale, rank, and often foul food, so that they could not maintain their health above two years." This meant that animals were bought into the city "at their third or fourth calving, when their yield was at its peak," kept for one season, and then "withdrawn."[92] The arrival of the animal at a later stage of development could be the reason why Rile was not yet in possession of his cow when he came to write his will: perhaps his father was raising her elsewhere for him. If these were the circumstances in which some of those Londoners who had them dwelled with their cows, then it is clear that this was not the long-term relationship between a few humans and a few animals that might be found on a rural smallholding. Instead, in the city kine would arrive as strangers and would not stay for long. The turnaround of stock would be swift, which would, inevitably, impact on the nature of the relationship with the animal. A cow kept for a season in a shed, as opposed to one living close by for over a decade, was likely to be regarded as a source of protein or income as much as, and perhaps more than, a colleague. For this reason, the attention given to such an animal was likely to have been different from what her country cousins might receive. Even while a London owner would still be attentive (such animals remained valuable and importantly productive, after all), the lack of long-term engagement might count against the cows, and this could affect the animal's physical well-being: knowing a transient animal's "tendency to suffer from certain disorders and how they responded to treatments in the past" might not be so easy.[93]

But it could also be that the short-term nature of the engagement between people and their cows in the capital meant that the link between instrumental and emotional attachment that made up household-family relations might have been broken and that Londoners' animals would have been looked after but would have been less likely to be regarded as affined to the humans they lived alongside than animals in rural communities. At the end of the shed cows' lives they would be, in Kerridge's euphemism, "withdrawn." Not

---

this instance, the tenement was divided and the kitchen was knocked down and a new building was erected in its place. "Book of Depositions by inhabitants of the parish and others in the course of a law suit (Johson vs Wright), relating to tithes and tenths issuing from the Black Mule in Fleet Street," LMA, P69/DUN2/B/030/MS03781 (1624).

92. Kerridge, *The Agricultural Revolution*, p. 178.

93. Louise Hill Curth, *The Care of Brute Beasts: A Social and Cultural History of Veterinary Medicine in Early Modern England* (Leiden: Brill, 2010), p. 65.

wishing to be wasteful, he writes, London owners would "throw some fat on them as quickly as possible by forcing them on brewers' grains, bean shells and cabbage leaves, and sell them off as cow-beef."[94] The process, of course, was the same as it would have been in the countryside; where sale for slaughter, then butchery and consumption awaited most aged dairy animals. What was different in the capital, aside from the living conditions and the speed of the shift from milk cow to cow beef, was the worldview that included the animal. All quick cattle, as we have seen, were movable goods, but the slaughter of an animal that was a short-term, functional, fungible presence was likely to trouble the owner less than that of a named, long-term resident. In this context, the "naturall reason" that might link people with their animals was perhaps more likely to be absent or more difficult to be brought to bear,[95] and the sense of the cow's kindness (of her generosity and recognizability) would have declined.

There would undoubtedly have been more than the small number of cows being used for milk production in the city than the wills show, and this sense of the increasingly transient nature of animals' participation in human life supports the lack of specific animal bequests in London wills. Where the testators of Essex only infrequently included pigs in their wills, despite the fact that such short-term residents were likely to be present in many households, so attentive identification in wills of the quick cattle of London might have been unlikely and the bequeathing of such animals might have been done without specificity—in collecting phrases—because of the temporary nature of their presence.

But there is another reason to recognize the persistent presence of livestock despite the evidence of the wills. It would be simplistic to view the shift from agriculture to other trades and the increasing buying in rather than farming of meat and dairy products as removing quick cattle completely from the experiences of the capital's residents. Instead, it is Londoners' changed encounters with them, not their disappearance, that should be emphasized.[96] Feeding "kine, goats, hogs, or any kind of poultry, in the open streets" of the

---

94. Kerridge, *The Agricultural Revolution*, p. 178.

95. Edmund Southerne, *A Treatise Concerning the right vse and ordering of Bees* (London: Thomas Orwin, 1593), C4r.

96. Claims that nineteenth-century urbanization led to a disappearance of animals, by Keith Thomas and John Berger, for example, have been challenged by Diana Donald and Jonathan Burt. See Thomas, *Man and the Natural World: Changing Attitudes in England 1500–1800* (London: Penguin, 1983), p. 181; Berger, *About Looking* (London: Writers and Readers Publishing Cooperative, 1980), p. 1; Donald, "'Beastly Sights': The Treatment of Animals as a Moral Theme in Representations of London c. 1820–1850," *Art History* 22, no. 4 (1999): 515–16; and Burt, "John Berger's "'Why Look at Animals?': A Close Reading," *Worldviews: Environment, Culture, Religion* 9, no. 2 (2005): 203–18.

capital, for example, was not banned until 1671, a fact that hints that such animals must have been present there before then, even if they left little mark in Londoners' wills.[97] But the poultry would not have been found in the street called Poultry by the end of the sixteenth century: by that time, what Chris R. Kyle has called the "onomastically transparent medieval city" had gone;[98] as Stow noted in 1598, *"Powlters* of late remoued out of the *Powltry* betwixt the Stockes and great Conduite in Cheape into *Grasse-streete* and S. *Nicholas* Shambles."[99] That is, they had moved from the east end of Cheapside to the west. This reference to St Nicholas Shambles takes us to a part of the capital where quick cattle would have been highly visible. The butchers of the west of the city could be found in this area (to the east they were to be found in The Stocks, Leadenhall, and Eastcheap), and this was therefore one of the places where the production of meat for the capital's growing population—also known as slaughter—was undertaken.[100]

Up until the early sixteenth century, the killing of animals in St Nicholas Shambles happened in the street, but this was prohibited in 1516, and by the early seventeenth century slaughterhouses had been built behind the butchers' shops.[101] The proximity of St Nicholas Shambles to Smithfield, the city's

---

97. See John Noorthouck, "Appendix: Act of Common Council for Paving and Cleansing the Streets (1671)," in *A New History of London Including Westminster and Southwark* (London: R. Baldwin, 1773), pp. 850–55, http://www.british-history.ac.uk/no-series/new-history-london/pp850-855, accessed 29 September 2016. In addition to livestock, Londoners would have encountered cats and dogs (as evidenced in the massacring of them during plague outbreaks), rats, and increasing numbers of horses. The number of carts licensed for carrying within the capital by city authorities grew from 40 in 1512 to 400 in 1580. See Mark S. R. Jenner, "The Great Dog Massacre," in *Fear in Early Modern Society*, edited by William G. Naphy and Penny Roberts (Manchester: Manchester University Press, 1997), p. 49; J. A. Chartres, "Road Carrying in England in the Seventeenth Century: Myth and Reality," *The Economic History Review* 30, no. 1 (1977): 73–94; Julian Allen and Michael Browne, "Road Freight Transport to, from, and within London," *The London Journal* 39, no. 1 (2004): 64–65.

98. Chris R. Kyle, "Afterword: Re-Mapping London," *Huntington Library Quarterly* 71, no. 1 (2008): 247.

99. Stow, *A Svrvay of London*, p. 63. Poulterers would also have sold rabbits, game, waterfowl, live pigs, bacon, brawn (boar's flesh), eggs, and butter. Ian Archer, Caroline Barron, and Vanessa Harding, eds., *Hugh Alley's Caveat: The Markets of London in 1598* (London: London Topographical Society, 1988), p. 10.

100. For an overview of the history of London meat markets, see Derrick Rixson, *The History of Meat Trading* (Nottingham: Nottingham University Press, 2000), pp. 133–36, 190–93.

101. Archer, Barron, and Harding, *Hugh Alley's Caveat*, p. 92; Rixson, *The History of Meat Trading*, p. 190. It is perhaps more than coincidental that Thomas More's *Utopia* was first printed in Latin in 1516, the same year that butchery was prohibited in the streets of London. In that work, More presented an ideal society in which all that might disrupt order was excluded, including all private ownership. In addition to having a communal culture, the Utopians relegated the work of butchery to slaves, a group made up of criminals, prisoners, and penniless foreigners. The slaughter of animals took place in specially constructed buildings "outside the city, where running water can carry away all the blood

live animal market, perhaps explains its continuing presence so close to Cheapside, an expensive residential and commercial neighborhood, but this closeness created inevitable problems.[102] The difficulty of ridding the street of the waste matter of slaughter and butchery was a problem from the Shambles's earliest days, and by the beginning of the sixteenth century the butchers, "at considerable expense, had made great vaults and underground drains, with a current of water to carry away offensive matter." In addition—and hinting at the incomplete success of the underground drains—Philip E. Jones records that in 1531 "an assessment was made upon the inhabitants of St Nicholas Shambles for the cleansing from time to time of the . . . gutters on either side of the shambles."[103] It is perhaps no surprise that even in 1612 Ben Jonson could depict the sewers of Fleet Lane—one of the thoroughfares that joined St Nicholas Shambles to the Fleet River—as running in "grease, and hair of measled hogs."[104]

Jones reports a different method being used to cleanse the surroundings in Eastcheap, one of the streets where butchers were to be found in the east of the city. Barrows of waste matter from the slaughter and butchery of animals were left out for collection overnight, when they were taken to be dumped in the Thames by Queenhithe. In the 1620s and 1630s, however, records exist of fines being issued to butchers who brought their tubs of entrails out too early in the evening or left them too far out in the street.[105] There was evidently no desire in London to "make [the slaughter] present againe" through its material presence (or even its material remains), as John Paget might have

---

and refuse." The narrator of the text, Raphael Hythloday goes on: "Bondsmen [i.e., slaves] do the slaughtering and cleaning in these places: citizens are not allowed to do such work. The Utopians feel that slaughtering our fellow creatures gradually destroys the sense of compassion, the finest sentiment of which our human nature is capable." For More, community—the sharing of goods as well as friendship—can happen only when the killing of animals Utopians have raised is done by others because such killing is the ultimate breach of compassion. More, *Utopia*, edited by George M. Logan and Robert M. Adams (Cambridge: Cambridge University Press, 1989), pp. 56, 107, 80, 57.

102. See Harding, "Families and Housing in Seventeenth-Century London," p. 121; and Vanessa Harding, "Cheapside: Commerce and Commemoration," *Huntington Library Quarterly* 71, no. 1 (2008): 77–96. London was not the only place where butchers' waste was an issue, of course. But the size of the capital meant that it was difficult to house all butchers near flowing water. In Maldon in Essex, for example, the row of butchers shops were on the High Street but were separated from the river Chelmer only by pasture and the town butts (the field for archery practice). See W. J. Petchey, *A Prospect of Maldon, 1500–1689* (Chelmsford: Essex Record Office Publication no.113, 1991), pp. 85–87.

103. Philip E. Jones, *The Butchers of London: A History of the Worshipful Company of Butchers of the City of London* (London: Secker and Warburg, 1976), pp. 81–82. A shamble was a table used to sell meat. *Oxford English Dictionary*, s.v. "shamble."

104. Ben Jonson, "Epigrams," in *Ben Jonson: The Complete Poems*, edited by George Parfitt (London: Penguin, 1988), l.145.

105. Jones, *The Butchers of London*, p. 84.

hoped.[106] Instead, the wish was clearly for the slaughter to be removed from sight. In such a context, killing animals would appear to be simply a worldly act from which otherworldly meaning was being lost. Slaughter was literally and figuratively a matter of the flesh.

But sight was not the only sense that would have been impacted by the presence of killing animals in early modern London, and other unwanted by-products were even less easy to control than actual entrails. Writing of St Leonard's, one of the parishes along Eastcheap, Paul Seaver notes that "73 houses crowded into [this] parish of less than one and a half acres. In the years at the beginning of the century when Nehemiah [Wallington] was growing up, 27 of his neighbors were butchers."[107] Some of the seventy-three houses may have been inhabited by more than one family, but the proportion of butchers was still very high. And even though butchery no longer took place in the street and thus may have been invisible to residents on the main thoroughfare, the smell would have continued to circulate: whatever the desire to shield oneself from these odors, they would have seeped into daily life.[108] In addition, the noise of animals being led to slaughter would also have been present and, to use Bruce R. Smith's phrase, would have limited the "acoustic horizon" of the neighborhood. In the countryside, after gunshot and thunderclaps, the bark of a large dog might have been the loudest thing encountered in this period, in and around numerous areas of London, the overwhelming noise might well have been that of livestock being driven, penned, and killed.[109]

And their presence was constant. In the 1603 edition of his *Suruay*, John Stow noted that there were "sixe score" (120) butchers in the city of London in 1533, and that "euerie one killed 6. Oxen a peece weekely, which is in forty sixe weekes, 33120. Oxen or 720 Oxen weekly."[110] The population of the city in the 1530s was approximately 50,000; by 1625 it was over 200,000. Using Stow's

---

106. John Paget, *Meditations of Death Wherein a Christian Is Taught How to Remember and Prepare for His Latter End* (Dort: Henry Ash, 1639), p. 52.

107. Seaver, *Wallington's World*, p. 131.

108. As Eleanor Margolies has noted, "Smells pose a pungent challenge. . . . [They] do not remain attached to their source, nor respect boundaries." Margolies, "Vagueness Gridlocked: A Map of the Smells of New York," in *The Smell Culture Reader*, edited by Jim Drobnick (Oxford and New York: Berg, 2006), p. 112. The "noisome stench" of Smithfield is a focus of Holly Dugan's reading of Ben Jonson's *Bartholomew Fair* (1614), a play that centers on the consumption of pork. Dugan, "'As Dirty as Smithfield and as Stinking Every Whit': The Smell of the Hope Theatre," in *Shakespeare's Theatres and the Effects of Performance*, edited by Farah Karim Cooper and Tiffany Stern (London: Bloomsbury, 2013), pp. 195–213.

109. Bruce R. Smith, *The Acoustic World of Early Modern England: Attending to the O-Factor* (Chicago: Chicago University Press, 1999), pp. 51, 50.

110. John Stow, *A suruay of London* (London: Iohn Windet, 1603) p. 189. This passage did not appear in the original 1598 edition.

figures, and assuming that meat consumption remained the same per capita and that butchers kept up with demand, it is thus possible that over 132,000 oxen were killed in 1625 at the rate of nearly 3,000 per week—within the just over one square mile that made up the city.[111] These figures assume that all Londoners avoided meat during the six weeks of Lent, but in 1620 John Taylor suggested otherwise: some "Beefe-braining *Butchers*," he wrote with a comic flourish, "doe scout into Stables, Priuies, Sellers, Sir *Francis Drakes* ship at Detford, my Lord Mayors Barge, and diuers secret vnsuspected places, and there they make priuate Shambles with kil-calfe cruelty, and Sheepe-slaughtering murder, to the abuse of *Lent*."[112] And it was not only Lent that butchers were said to ignore; the Sabbath was also not sacred to them. The Butchers' Company Ordinances of 1607 had to forbid members from driving, killing, or selling animals on Sundays. But in spite of the ban, slaughter continued seven days a week, as did selling: in 1627, the Middlesex Sessions Rolls reported "the great and insufferable abuses and disorders, committed by poulterers and butchers in opening theire shopps and selling flesh and poultry wares upon the sabboth dayes both before and in tyme of divine service in the markettes kept at Cowcrosse, Smithfeildbarrs, St. Johnstreete, Feildlane, Eastsmithfeild, St. Katherins, Nortonfolgate, Wapping, Shorditch, Whitechappell and other places of this county."[113] The places listed are just outside the boundary of the city to the west, north, and east, and, as Margaret Dorey notes, documents reveal that the 1607 ordinances failed to change butchers' behavior within the boundary too.[114]

In addition to the noise and the stench of blood and entrails that the almost constant slaughter would have produced in the particular areas of the capital where butchers were based, many more Londoners would have had to contend with the frequent presence of live animals being driven through the streets to Smithfield from across the rest of England as well as from as far afield as Wales and Scotland. Cows, sheep, pigs, and poultry were sold at the market, which was just over a third of a mile to the north of St Paul's Cathedral and

---

111. Alessandro Magno wrote that "it is extraordinary to see the great quantity and quality of the meat—beef and mutton—that comes every day from the slaughter-houses in the city. . . . It is almost impossible to believe that they could eat so much meat in one city alone." Quote in Barron, Coleman, and Gobbi, "London Journal of Alessandro Magno," p. 143.

112. John Taylor, *Iack a Lent his beginning and entertainment* (London: I. T., 1620), A3r and B3v. This was brought to my attention in Margaret Dorey, "Controlling Corruption: Regulating Meat Consumption as a Preventative to Plague in Seventeenth-Century London," *Urban History* 36, no. 1 (2009): 30–31.

113. "Middlesex Sessions Rolls: 1627," in*Middlesex County* Records, vol. 3, *1625–67*, edited by John Cordy Jeaffreson (London, 1888), pp. 13–17, http://www.british-history.ac.uk/middx-county-records/vol3/pp13-17, accessed 11 October 2016. A similar complaint had been made in 1622.

114. Dorey, "Controlling Corruption," p. 30.

was held on Mondays and, from 1613, Wednesdays—the Wednesday market added because of the population increase.[115] Herds of fattened animals were driven into London along the capital's main thoroughfares,[116] and while the drovers stayed in many of the city's inns, the quick cattle were kept in pens at Smithfield from their time of arrival until the market took place. According to Alec Forshaw and Theo Bergström, "there were pens and rails for about 4,000 cattle, sheep and pigs."[117] After they were sold, the animals were taken from the market to their final destinations, and for some the journey was not long. In 1637, nine slaughterhouses were located in the ward of Farringdon Within near Smithfield.[118]

Nearly fifty people lived above these slaughterhouses, and the proximity of human life to animal death that is evidenced in such close cohabitation can be read as an extreme version of what many Londoners experienced on a regular basis. Where cohabitation in the countryside might have been with the same cow over a period of years, this was improbable in the city. Instead, in London contact was more likely to be brief, smelly, noisy, and disruptive. The activity of slaughter also took place in the towns and villages of Essex, of course, but the scale and the fact that this may have been the main aspect of animal husbandry that was visible (or olfactible or audible) to the inhabitants of parts of London must signal a transformation in engagement. If Gervase Markham's husbandman began and ended his day by feeding his animals, Nehemiah Wallington, the Eastcheap wood turner, might have begun and ended his smelling the production of meat. And during the day, while the countryman could hear what Nicholas Breton called the "musique" of "the Cowe lowing, the Eue bleating, & the Foale neighing," Wallington's hand-turned lathe might have been unable to drown out the sound of frightened animals.[119]

---

115. Archer, Barron, and Harding, *Hugh Alley's Caveat*, p. 12. In addition, according to John Stow, "euery fryday (vnlesse it bee a solemne bidden holy day) is a notable shew of horses to bee sold." Stow, *A Svrvay of London*, p. 61.

116. J. A. Chartres writes that "broadly speaking, routes into Kent, Sussex, Surrey, and parts of Hampshire ran through Southwark to [London Bridge]; Fleet Street and the Strand served the bulk of the two south-western roads; traffic to and from the north and east used Bishopsgate; the Oxford road led into Holborn; and the north west, Yorkshire, and the midlands was served by Aldersgate Street, or from Smithfield through St John's Street." Chartres, "The Capital's Provincial Eyes: London's Inns in the Early Eighteenth Century," *The London Journal* 3, no. 1 (1977): 29.

117. Alec Forshaw and Theo Bergström, *Smithfield Past and Present* (London: Heinemann, 1980), pp. 20, 36. They note that "Monday morning was the busiest. . . . Throughout the Sunday night a great stream of well-fattened animals plodded down St. John Street from the fields and lairs of Islington." (p. 36).

118. Figure cited in Baer, "Housing for the Lesser Sort in Stuart London," p. 80.

119. Nicholas Breton, *The Court and Country, Or A Brief Discourse Dialogue-Wise Set Down between a Courtier and a Countryman* (London: G. E. for John Wright, 1618), B2v.

What this would have meant is worth contemplating. If, as the figures seem to show, the human population of London was shifting and growing throughout the brief period that is the focus of this book, then it must have included many who had grown up alongside, worked with, slept in rooms above, quick cattle. A child raised on a smallholding might have chased the poultry in the yard, herded the pigs, talked to the cow as part of their daily routine. They might even have had a lamb of their own to keep tabs on. In London, such interactions were unlikely. This must have had an impact on that child when he or she grew up and moved to the capital to try its opportunities. Donald Lupton stated that the countryman "many times looseth himselfe" in the city,[120] and we might now wonder whether that loss might also have been in terms of relationships with quick cattle—creatures who were regarded as only a little less than kin in the country. And for many who attended the theater in the capital, the experience of watching Lodge and Greene's comic representation of one character's devastation at having "No Cow" would, perhaps, have been tinged with compassion in a way that is now incomprehensible to us.[121]

We can see, therefore, that in London the understanding of the household-family was being transformed in ways that went beyond overcrowding, possible separation from community networks, or the change in the concept of privacy that poor housing created—although they would certainly have been significant changes for many. It was also transformed in relation to how animals were felt to fit into existing categories. In Essex, as the wills show, quick cattle were property but they were also a means of reinforcing family ties (whether in life or in death, in a loan or a bequest), and some of them were long-term co-workers with names. In London, in contrast, whether in their absence from the purview of the pater familias in many, perhaps most, of the capital's households or in their only short-term, fungible presence in others, these new living arrangements would change the engagements people had with quick cattle and that would affect those people and the animals they encountered. No longer members of one's own household-family, people might associate the cows and sheep being driven to Smithfield or brought to the butcher's shop, the horses being whipped along the increasingly crowded streets, and the pigs and poultry wandering around not with collaboration, productivity, and the (meaningful) discomfort that attended that but with filth, noise, and overcrowding. And their kindness—the sense of their being affined to their human colleagues but also of their being active, generous

---

120. Lupton, *London and the countrey carbonadoed*, p. 2.
121. Lodge and Greene, *A Looking Glass for London*, B4r.

participants in their productivity—would have declined. The animals being driven through the streets of London would not be invited to display preferences about music or the clothing of those they worked with, and if they were regarded as possessing their own moral codes those codes were likely to be an annoyance rather than a guide. Instead, in the capital, care was probably universal, brief, and utilitarian and conversation across species boundaries one way. And this meant that the physical closeness of animals in London was uncomfortable rather than comforting. The sense of kindness was changed.

But alongside the migrants who might have felt the loss of close contact with quick cattle in the capital, there were those like Nehemiah Wallington who were born and raised in London and so had never had the experience of animals that their country-born neighbors might have had. For people such as him, perhaps, the driving and slaughter of animals was the norm and closer encounters with quick cattle potentially frightening or unpleasant. This is a perspective in which human kindness to and with animals is, once again, of a different quality. Just as independent milkmaids might have to deal with many and individualize few, so looking away from rather than being attentive to animals might have been the daily response of some people to the creatures they encountered. And thus, even in their very presence, the quick cattle of London might have become increasingly invisible as sentient—affined—creatures to many who they passed in the streets. Unkind-ing and unkindness were co-constitutive, as Thomas Stevens knew.

But the detachment of people from animals in London was not complete. Although quick cattle were perhaps beginning to be reduced to the status of technology, they were always uncomfortably responsive machines, fleshy things that challenged the boundary between subject and object. And Shakespeare's image of the relationship of the cow to her calf as the latter is taken away by the butcher reminds us in a different way than Lodge and Greene's play that the capital's theater audiences still understood the potential for empathy, still possessed the sense that another way of engaging was possible. Perhaps King Henry's extended simile should be read as a reminder that the feeling of interspecies kindness was often experienced in the breach in this period. The loss contained in the image, you might say, is not only of a cow for her calf or of a royal master for his faithful advisor; it is also of a population for a world view in which human lives were lived out on the basis of connection, analogy, and kindness with animals.[122]

---

122. I have written elsewhere about the role bear baiting played in confirming concepts of humanness in the capital and this is perhaps linked here. As they disassociated themselves from live-

These changes were not only found within the heart of the city, however, and the size of London is reflected in the fact that the impact of urbanization on people's relationships with quick cattle exceeds its boundaries. Almost exactly half of all the Middlesex wills in the LMA dataset (442 of 883) are by testators from parishes that are outside the city of London but within two miles of St Paul's Cathedral—this includes Stepney and Whitechapel to the east, Clerkenwell and Islington to the north, and St Giles in the Fields and Westminster to the west. Only two of those 442 testators—less than 0.5 percent—included specific animal bequests in their will.[123] Edward Purlock, a husbandman from Islington, bequeathed his children "my five kine ^wch are in my Backside^ & my horse"—perhaps evidencing another small dairy near to the capital,[124] and William Lucas, an ostler from St Giles in the Fields, left his brother John Sodam "2 mares one sorrell ^standing at the Angell wthout temple bar in St Clement Danes in the county of Midds,^ the other a browne black standing . . . ^at the Antelopp in Holborne in the pishe of St Anddrowes."[125] In addition, two testators—William Thomas, a yeoman,[126] and Katherine Johnson, a widow,[127] both from Cowcross, just north of Smithfield—referred to quick cattle in their will's collecting phrase.[128] It is impossible to know widow Johnson's occupation, of course, but the fact that Purlock was only one of two husbandmen[129] and Lucas the only ostler within two miles of London in the LMA dataset suggests that what is crucial here is what this tells us about who wrote a will rather than who owned animals—that using these documents as evidence skews the picture because will writing was not the norm for many in the parishes neighboring the city. The fact that Thomas is the only one of thirty-two Middlesex yeomen from within two miles of the capital who included any reference to his animals in his will is less easy to explain,

stock, people sought new ways to comprehend their own species. See *Perceiving Animals: Humans and Beasts in Early Modern English Culture* (Basingstoke: Macmillan, 2000), pp. 11–33.

123. The very low proportion of testators from inner Middlesex who included specific animal bequests in their will may be due to the high number of testators who were sailors and mariners in riverside areas such as Wapping, Ratcliffe and Limehouse. None of the testators in this group included a specific animal bequest in his will. However, when I excluded wills by seagoing testators (129 in total), the proportion of Middlesex wills by testators who lived within two miles of St Paul's Cathedral that included specific animal bequests increased only to 0.6 percent.

124. LMA, DL/C/B/008/MS09172/031 217 (1620), made probate 48 days after writing.

125. LMA, DL/C/B/008/MS09172/031 262 (1620), made probate 6 days after writing.

126. LMA, DL/C/B/008/MS09172/039 298 (1630), made probate 585 days after writing.

127. LMA, DL/C/B/008/MS09172/042 65 (1634), made probate 815 days after writing.

128. The fact that both wills were written over a year before their date of probate could have influenced the testators' use of the general phrase "quick cattle."

129. The other husbandman is LMA, DL/C/B/008/MS09172/035 27 (1625), made probate 218 days after writing. No animals are mentioned in the will and the collecting phrase refers only to "All the Rest of my goods and Chattells not yet geuen & bequeathed."

however.[130] In the large ERO dataset nearly 15 percent of yeomen included specific animal bequests, but here the figure is 3 percent. One reason for this statistical difference might be the possibility that for those who lived close to London who referred to themselves as yeomen, their engagement with their animals was more short-term, meaning that those animals were less likely to be specified in wills: the lack of specific animal bequests here evidencing the transient nature of human-livestock relations close to the city, and underlining the particular value, once again, of a will's collecting phrase.

The preponderance of bequests of animals from Middlesex towns and villages located farther from St Paul's Cathedral, however, seems to reveal a return to something that looks familiar and appears to mark the end of the urban spread of the capital in the early seventeenth century in a particular way. Where in Essex specific animal bequests were found in 10 percent of all wills, in the 441 Middlesex wills of people who lived more than two miles from St Paul's, 15 percent (66 wills) include specific animal bequests, with just under 11 percent (31 of 293) in those from between two and nine miles from the city and over 23 percent (35 of 148) in Middlesex parishes located ten or more miles from London.[131] The increased proportion of Middlesex testators who lived more than ten miles from the capital and mentioned animals in their wills seems to reflect John Norden's assessment of the county. He wrote in 1618 of husbandmen living in the "body or heart" of Middlesex that they are "so furnished with kyne that the wife or twice or thrice a weeke conveyeth to London mylke, butter, cheese, apples, peares, frumentye, hens, chyckens, egges, baken, and a thousand other country drugges, which good huswifes can frame and find to gett a pennye. And this yeldeth them a lardge comfort and releefe."[132] The significant income that would have come from such sales might, in turn, have meant that testators were more likely to recognize the particular importance of animals to the household-family's well-being and that that was reflected in their specifying of quick cattle in their wills.[133]

---

130. Yeomen made up 7 percent of testators who lived within two miles of St Paul's; in the large ERO dataset they made up 25 percent.

131. Overall, of the 66 wills that included animals from outer Middlesex, 19 (20 percent) included horses only. In wills by testators from within the City of London that figure was 54 percent, whereas in the large ERO dataset it was 9 percent. For wills from Middlesex parishes ten and more miles from the city, the figure was 23 percent (8 wills). In the Middlesex parishes two to nine miles from London, it was 35 percent (11 wills).

132. John Norden, *Surveyor's Dialogue* (1618), quoted in Fisher, "London Food Market," p. 55. Frumenty is cracked wheat.

133. This might also be the case with the wills from Essex parishes under the jurisdiction of the Diocese of London that also include a slightly higher than average number of specific animal bequests: 22 of 179 (12 percent) wills include them. These parishes are in the southwest of the Essex—as far north as Roydon and Great Parndon (each of which is nineteen miles from the city) and going south

However, despite the potential for financial gains to be made by selling to the hungry capital, on the whole the wills by testators from parishes more than two miles from London in the LMA dataset are very similar to those from Essex in the large ERO dataset. We can find in them, for example, no evidence of large herds of animals (the greatest number of cows mentioned in any will is nine). Indeed, we can see in the wills that do include specific animal bequests evidence of a world that Margery Reade and her neighbors in West Bergholt might recognize: there is a cow called Bullhead in Ealing (ten miles from the city),[134] lambs are given to children in Edmonton (seven miles from London) and Enfield (ten miles),[135] and a testator in Acton (eight miles away) left the income from "eight stocks of bees" to be used "for the clothinge of my sonn William Childe till such time as he goe to be an Apprentice," a bequest that might be read as linking the exemplary value of bees with a boy's early education.[136] The fact that the wills from outer Middlesex are similar to those from Essex raises the possibility, not that the impact of the capital was not manifested in the Middlesex wills, but that the presence of London was just as visible in the wills of many of the testators from Essex, that the city's influence was felt across a wide area of the surrounding countryside. In addition, the similarity in the wills might imply that even where the potential market for live or dead animal products was substantial, that did not mean that quick cattle were viewed simply as productive things or fleshy machines on smallholdings. The wills in both datasets show that long-term, day-to-day closeness to animals appears to breed a particular kind of attention, whatever the financial incentives. Indeed, one Middlesex village offers a useful illustration of the complexity of tracing the shift from rural to urban; and the impact of the urban on the rural in wills in this period.

Tottenham, a parish six miles north of St Paul's Cathedral, is on the Old North Road, one of the key drovers' routes into the capital, and records show that arable farming had been replaced as the central plank of its economy by the fifteenth century, when numerous London butchers bought up pasture there to use for grazing the animals they had purchased from farther afield. Their intention was to fatten up the animals close to the point of sale and thus

---

through Nazeing, Epping, Waltham Holy Cross, Loughton, Chingford, Woodford, Walthamstow, and Leyton and east as far as East Ham and Barking (seven and eight miles from London, respectively). Wills from some of these parishes (Roydon, Great Parndon, Nazeing, Chingford, Leyton, East Ham, and Barking) are also found in the large ERO dataset.

134. This is in the will of the yeoman Robert Maynerd; LMA, DL/C/B/008/MS09172/038 118 (1627), made probate 32 days after writing.

135. LMA, DL/C/B/008/MS09172/038 79 (1627), made probate within 434 days of writing; LMA, DL/C/B/008/MS09172/040 35 (1631), made probate 30 days after writing.

136. LMA, DL/C/B/008/MS09172/039 191 (1630), made probate 215 days after writing.

make the best price when they sold them at Smithfield.[137] Diocese of London wills from the parish that were proved in the period 1620 to 1634 seem to reveal very little evidence of this practice, however. Not only do they lack evidence of substantial property ownership,[138] the business of fattening up is not visible in testators' bequests of movable goods either, despite the history of butchers buying up land in the parish. The logic of previous arguments would, of course, allow that the short-term presence of animals for fattening up might make it unlikely that they would be specified in a will and that the invisibility of quick cattle in those documents is evidence of the transient nature of relationships with them. That may, of course, be the case here. But such (invisible) transience is absent in a couple of the Tottenham wills that, in fact, make visible a world of relationships with animals that is reminiscent to that found in wills in the large ERO dataset. If nothing else, these two Tottenham wills show that small-scale enterprises persisted alongside large-scale fattening up.

The 1625 will of the husbandman John King is one of two wills from the parish that include specific animal bequests (12 percent, although a third will includes reference to "hay corne cattell householdstuffe").[139] As such, the proportion of testators in Tottenham who mentioned animals is slightly higher than for other parishes located between two and nine miles from the capital, where it is just under 11 percent. In his will, along with the lease of his house, which he left to his wife Joan, entailing it to his oldest son John after her death, King left that son a "brended Cow" and in addition he bequeathed to John, Mary and Elizabeth, the three children of his brother Thomas "thre of my best Ewe lambs," and to Roberta Watkins (perhaps a servant) "one of my sheep."[140] As such, this will is like many from rural Essex: it seems to reflect the existence of small-scale production, it evidences close kin networks, and shows the potential emblematic value of lambs. King's animals were more than simply flesh.

A similar picture emerges in the other Tottenham will that includes specific animal bequests. The 1624 document of the yeoman Robert Morris

---

137. Douglas Moss, "The Economic Development of a Middlesex Village," *Agricultural History Review* 28, no. 2 (1980): 113–14.

138. Only seven of seventeen Tottenham wills in the LMA dataset (41 percent) include a reference to real estate. In the large ERO dataset that figure is almost 45 percent. One of those seven wills included property in St Giles without Cripplegate in the city; LMA, DL/C/B/008/MS09172/040 153 (1631), made probate 20 days after writing. The property in this will was left to the parish for the support of the Tottenham poor. When this will is excluded, the proportion of Tottenham wills that included real estate falls to 35 percent.

139. LMA, DL/C/B/008/MS09172/038 212 (1625), made probate 152 days after writing.

140. LMA, DL/C/B/008/MS09172/036 148 (1625), made probate 124 days after writing.

includes reference to more substantial landholdings than are evident in other wills from the village. It notes that an unspecified amount of land had already been given to Morris's son John and then refers to a field in Edmonton (a mile from Tottenham) and two houses, one "by woddgreene" (just over a mile from Tottenham) and the other "at tottenham streete."[141] The detail is missing, but it is possible that the land given to John or the field in Edmonton were used for fattening animals. But, even if this is the case (and the will does not make it clear) we can see something else in the will, too, if we look at Morris' other possessions. Beyond including bequests of money amounting to over £60, it presents a world that would have been familiar to Christopher Fuller, who lived forty-five miles northeast of Tottenham. Morris bequeathed to his son Lawrence "the wenell calfe" and "the Kowe wch I call flemishe" and to another son, Richard, a "Kowe called Starre." A third son, William, who received the lease of the field in Edmonton, also got a "Kowe called Coxhead," and Morris's wife Ann received "Six of my beasts" and "all the geese & my hoggs & the henns & the horsses."[142] If this accounts for all the animals he possessed at his death, then Morris's will makes visible a good-sized family farm. The inclusion of one calf perhaps shows that he was also involved in production for the market, an inevitable consequence of owning and milking at least nine cows. As we have seen, cows were likely to calve in the first few months of the calendar year and calves were usually weaned in May, thus, Morris's will, written on 15 October, might include the last of the year's weanels that the household was selling for meat. Or perhaps it was one that was being kept longer to increase its size before sale, or to be an addition to his dairy herd. Interestingly, Morris also left his sons Lawrence and John, who were named as his executors, "my yoke of Oxen towards the payinge of my debts."[143] This is one of only two references to oxen in any will from this period in either London, Middlesex, or Essex,[144] and it might be that Morris directed that these animals be sold to cover his debts because he knew that his

---

141. LMA, DL/C/B/008/MS09172/034 134 (1624), made probate 24 days after writing.

142. LMA, DL/C/B/008/MS09172/034 134.

143. LMA, DL/C/B/008/MS09172/034 134. One other Middlesex will includes a named cow. The Yeoman Robert Maynerd of Ealing bequeathed his wife Sarah "One blacke Cowe Called Bullhead and one other Redd brinded Cowe." LMA, DL/C/B/008/MS09172/038 118.

144. The other is in the will of Henry Boomer of South Mimms, which is fourteen miles northeast of London. In addition to the mare and dung cart Boomer left his son Sammell, he bequeathed him "two oxe yoakes [and] one paire of draughts." LMA, DL/C/B/008/MS09172/034 124 (1624), made probate 35 days after writing. I take the "paire of draughts" to be the animals, a usage the OED allows for: Oxford English Dictionary, s.v. "draught." Christopher Fuller's "two 2 yeare old steares" may also have been draft animals; see ERO, D/ACW 8/270 (1620), made probate 71 days after writing.

more up-to-date children would not want to use them, as by this time oxen had become outmoded; horses had replaced them for traction purposes. As such, the bequest might be similar to Mary Archer's wish that her "two old cowes" be sold.[145] Like the question of whether his land was used for fattening up and the fate of the weanel, however, this interpretation of Morris's yoke of oxen is uncertain. But the aspects of animal husbandry that are visible in his will—the presence of the weanel and the oxen—seem to show the simultaneous existence of the commercial and the old-fashioned in the parish. What is clear in the document is that three cows (at least) have names, something that reveals his closeness to these animals and the longevity of his relationship with them. Naming an animal is unlikely to reflect the fast turnaround of dairy animals that might be found in the sheds of the capital. Like his neighbor John King's will, Robert Morris's seems to reflect the existence of a household-family of humans and animals in a village that had a history of being used by butchers.

However, one other Tottenham will is rather unlike these two, and not just because it includes no animals. This final document is the 1632 will of the gentleman Samuell Willcockes and I suggest that it might begin to reveal the existence of what was to become a new norm of animal agriculture. As in Thomas Stevens's story of his brother's treatment of him, the importance of closeness is raised through its absence in this document, for where in Essex and London proximity (with its differing implications) is key—in one place it breeds affinity, in the other discomfort—in Willcockes's dying wishes it is distance that is crucial.

In his will Willcockes left freehold and copyhold land and a house with gardens and orchards in Tottenham, "whearin Mr Manwood doth now dwell." In addition, he bequeathed lands and tenements with both copyhold and freehold land in Ipswich and Mickfield in Suffolk and lands with tenements and hereditaments in Shropshire. He also noted in his will that he wished to be buried where he had been born, in Bermondsey, on the south of the river Thames.[146] The fact that Willcockes possessed land in more than one place and no longer lived in Tottenham could suggest that he had come from a family that originated in Shropshire (where his hereditaments—inherited property—were located) and had moved to Bermondsey, where he had been born. It might also show that, in addition to possessing his inherited lands in Shropshire, he had decided to buy land in Suffolk and land and property in Tottenham because both places offered the prospect of profitable trade

---

145. ERO, D/ABW 43/197 (1620), made probate 107 days after writing.
146. LMA, DL/C/B/008/MS09172/040 257 (1632), made probate 25 days after writing.

with the capital. His Shropshire pastures might, indeed, have supplied the animals for his herds in Suffolk, which, in turn, supplied cheese to the London market,[147] and might also have provided the quick cattle that were driven down to Tottenham for fattening up in the final days before they were sold in the capital.[148] But no quick cattle are mentioned in Willcockes's will and its collecting phrase refers simply to his "goods & Chattells."[149] The geographical dislocation that is evident in Willcockes's life and possessions can be linked, I suggest, to this absence of quick cattle in his will. Where Margery Reade could count her animals (four sheep and five lambs) and could judge that it was the "blacke Cowe" who should be lent to Elizabeth rather than, say, the "litle Cowe," and where Christopher Fuller could distinguish between "my brended Cowe, my litle black northen Cowe & my Cowe called Shaler," Willcockes, residing at a distance, could not know the animals living on his land—certainly not individually and perhaps not even by quantity. In this context, the collecting phrase was particularly useful. It allowed for possessions to be transferred, but it also allowed for the fact that those possessions were unknown to their owner. To Samuell Willcockes, such animals were simply movable goods over which he had (as the law put it) absolute ownership.

The relationship between Willcockes and the animals he was the legal owner of (if it can be called a relationship) might, in fact, have been similar to that which Baron Petre had with the quick cattle in his possession at Thorndon Hall in Ingrave, in Essex (which is twenty miles northeast of London). Agricultural accounts from the 300-acre estate during the period of the large ERO and LMA datasets record the coming and going of "Cattell in the lower grounds in the charge of William Snowe," and the role of Snowe makes clear that the baron himself had little, if anything, to do with these animals.[150] But it is not just the fact that the owner of the land hired someone to work alongside the quick cattle that is notable (that is to be expected for a peer of the realm); it is the nature of the agricultural production at Thorndon Hall that is important (it is difficult to call it husbandry). However substantial the consumption of the household itself might have been, the estate was clearly not working just to provide its residents with meat: it was also a commercial

---

147. Suffolk was one of the capital's key suppliers in this period. See Fisher, "The Development of the London Food Market," p. 51.

148. For the place of Shropshire in the cattle trade from Wales to London, see Caroline Skeel, "The Cattle Trade between Wales and England from the Fifteenth to the Nineteenth Centuries," *Transactions of the Royal Historical Society* 9 (1926): 135–58.

149. LMA, DL/C/B/008/MS09172/040 257.

150. Thorndon Hall Accounts—Petre Family of Ingatestone," ERO, D/DP A52 (1633–1642). The land was "emparked" by license from Henry V. See "Thorndon Hall, Ingrave-Essex," http://www.thorndonhall.co.uk/thorndon-hall-history.htm, accessed 12 July 2017.

enterprise.[151] To take just a brief period of entries in the surviving records: in September 1633, Snowe is noted as having received eight "runts" (small oxen or cows) from Epping Fair, sixteen from Brentwood Fair, and six from Bushey Fair (twelve miles, two miles, and thirty miles from Ingrave, respectively). The following month he is recorded as having sold all the runts he bought in Epping to "Chalke the butcher." In January 1634, another eighty-one runts were received onto the estate (including sixty-one from Shropshire), and that June, twenty-two were sold to a John Fuller and the following month another fifty-four were sold to Chalke.[152] The accounts were necessary to give the Petre family an idea of what quick cattle they were in possession of as only Snowe, the manager, and his assistants would actually have engaged with the animals, and then only briefly. And the short-term nature of the animals' stay would mean that no sense of an individual runt's history would have existed. Indeed, at Thorndon Hall it is possible that the only animals that were individualized were dead ones. In the accounts for January 1634 is noted:

Runtes rec of Shropshire	lxi
Whereof died	i[153]

While very different, I am suggesting that Willcockes's will and the accounts at Thorndon Hall offer a sense that it was not only living in the capital that changed people's relationships with animals. Both documents seem to show that changing agricultural practices were having an impact too.

Cicely Howell's research in the Midlands allowed her to suggest that in the mid-seventeenth century there was a "parting of the way between

---

151. The "Very Detailed Weekly Provision Book" of Audley End Estate, Saffron Walden, home of the Cornwallis family, reveals how much meat a large house might consume. In September 1627, for example, the list of provisions included 483 pounds of beef plus pork, mutton, pigeon, and rabbit, "hare kild byye hauks and spaniells," and eight partridges "kild with the hauke." ERO, D/DBy A382 (1617–1629).

152. ERO, D/DP A52.

153. ERO, D/DP A52. In this context it is perhaps unsurprising that when Sir William, 2nd Baron Petre, the owner of Thorndon Hall during the period of the large ERO dataset, wrote his will (which was proved in the Prerogative Court of Canterbury), he included the following as his collecting phrase: "all my Plate silver vessells candlesticks of silver Leases debts money Beddes Bedding hangings Napery Lynnen, Linnen-cloth household stuffe, Horses Geldings Mares Coltes Oxen sheepe and other Cattle Stocke things or goods whatsoeuer not otherwise by me before or hereafter bequeathed." NA, PROB 11/174/394 (1637), made probate 1,625 days after writing. This is all valuable stuff. What is also unsurprising is how much more careful he was in the description of horses than other animals; these, perhaps, were the animals he might have had most direct contact with and expended the most money on without making money back. Peter Edward notes the Petre household's regular use of a farrier during the reign of Elizabeth I. Edwards, *Horse and Man in Early Modern England* (London: Hambledon Continuum, 2007), p. 64.

husbandman and yeoman . . . the one to remain the humble villager, the other to become the local squire."[154] Likewise, in his study "The Rise of Agrarian Capitalism and the Decline of Family Farming in England," Leigh Shaw-Taylor has argued that "family farms were probably of minor importance as early as 1600 and a residual element of the landscape" 100 years later. By 1800, he writes, "The agrarian landscape was dominated by a tripartite social structure in which most of the land was owned by large landowners, rented to large-scale tenant capitalist farmers, and worked by agricultural proletarians."[155] The brief period covered in this book does not allow for such shifts to be tracked in the parts of southeast England that have been a focus here, but the evidence that the wills and other records reveal of the lives of the people and animals of London and the agricultural landscapes of Essex and Middlesex seem to show that even in that fifteen-year period diverse practices were possible, and it is tempting to read this as evidencing the shift that historians have found more generally. Samuell Willcockes's experiences are certainly likely to have been very different from his neighbor John King's, for whom it is possible that lambs were emblematically meaningful, for example. And it might be that Robert Morris's will reflected his buying up of new properties, just as the Hertfordshire yeoman Robert Jacobb had bought £255 worth of real estate for his namesake son.[156]

It is worth contemplating, in the light of Howell and Shaw-Taylor's findings, what the future might hold for Morris's sons John, Lawrence, Richard, and William. But we should wonder what that future held for the offspring of Goldelocks, Flemishe, Starre, and Coxhead too; what the experience of the issue of the ewe lamb Richard Freebody left to "Ric Martin sonne of Ric Martin" that was "to be deliu'ed vnto him foure yeares hence if he live soe longe" was going to be,[157] because these changes would have affected them as well. Shaw-Taylor's conclusions imply that the lives of people and their animals that are made visible in the wills from early seventeenth-century Essex

---

154. Cicely Howell, "Peasant Inheritance Customs in the Midlands, 1280–1700," in *Family and Inheritance: Rural Society in Western Europe, 1200–1800*, edited by Jack Goody, Joan Thirsk, and E. P. Thompson (Cambridge: Cambridge University Press, 1976), p. 152.

155. Leigh Shaw-Taylor, "The Rise of Agrarian Capitalism and the Decline of Family Farming in England," *Economic History Review* 65, no. 1 (2012): 47, 26. See also Paul Glennie and Ian Whyte, "Towns in an Agrarian Economy 1540–1700," in *The Cambridge Urban History of Britain*, vol. 2, *1540–1840*, edited by Peter Clark (Cambridge: Cambridge University Press, 2000), p. 176; and Bruce M. S. Campbell and Mark Overton, "A New Perspective on Medieval and Early Modern Agriculture: Six Centuries of Norfolk Farming c. 1250–c. 1850," *Past and Present* 141 (1993): 38–105, esp. pp. 82–92.

156. ERO, D/AMW 3/172 (1617), made probate 47 days after writing.

157. ERO D/ABW 43/71 (1621), made probate 25 days after writing. A copy of this will is at ERO D/AEW 17/4 and I excluded it from the large ERO dataset because it is a copy. The existence of two versions in two different jurisdictions (ABW and AEW) tells us that Freebody had property in each.

and elsewhere and the particular concept of kindness that their lives were premised on were becoming endangered even at the time those testators were recording their dying wishes. In the emerging world, affection—which had been founded, as we have seen, on the inseparable economic and emotional value of the animals—was changing. Care was still being taken, of course (animals remained economically important), but affinity, the sense of being of a kind with them, was being undermined. Even as Margery Reade was lending her daughter a cow, the relationship of the commercial milkmaid to the cows she milked, you might say, was being replicated across animal agriculture. This not only impacted how affinity was experienced, however; it also changed how animals' kindness might be understood, with the animals themselves, perhaps, coming to be viewed less as generous co-workers and increasingly as productive things that engaged without minds in the processes of husbandry, as creatures whose participation might be assumed to be a mechanical response rather than evidence of collaboration. In addition, instrumentality and affection, like the connections between people in London, were being unmoored from each other as animals' place in the household-family changed, and kindness—caring for animals but also recognizing a kinship with them—was being eroded. The outcome, as Thomas Stevens and Hamlet knew, was likely to be a lowering of their status, an emotional distancing. Animals that had been perceived as "a little less than kin and more than kind" were being transformed into figures on a page: i dead, lx to go.

✌ AFTERWORD

Bovine Nostalgia

In 1688, a year after the first recorded appearance of the word livestock in English, the Swiss scholar Johannes Hofer described for the first time in the medical terminology of the day a disease that he recognized as affecting "certain youths" who, "unless they had been brought back to their native land[,] . . . met their last day on foreign shores." This potentially fatal disease Hofer termed "nostalgia." The word, he wrote, is "Greek in origin and indeed composed of two sounds, the one of which is *Nosos*, return to the native land; the other, *Algos*, signifies suffering or grief," and names "the sad mood originating from the desire for the return to one's native land." This mood, he noted, particularly affected "young people and adolescents" who, "when they are sent forth to foreign lands with alien customs, do not know how to accustom themselves to the manners of living nor to forget their mother's milk." The symptoms of the disease included "continued sadness, meditation only on the Fatherland, disturbed sleep either wakeful or continuous, decrease of strength, hunger, thirst, senses diminished, and cares or even palpitations of the heart, frequent sighs, also stupidity of the mind—attending to nothing hardly, other than the idea of the Fatherland."[1] Swiss mercenary soldiers—professional military men

---

1. Johannes Hofer, "Medical Dissertation on Nostalgia, 1688," translated by Caroline Kiser Anspach, *Bulletin of the Institute of the History of Science* 2 (1934), pp. 380, 381, 383, and 386.

who would serve foreign nations—were felt by medical practitioners to be particularly affected during their lengthy travels abroad and when, in 1710, Thomas Zwinger reprinted Hofer's text with amendments, he added to the list of things that might particularly affect them "a sweet melody of Switzerland which tends to produce homesickness in everyone who hears it." This melody was termed the "Kühe–Reyen" and was the tune that was played on the horn of an alpine herdsman as he drove his animals.[2] Cows were present almost at the origin of nostalgia.

Nostalgia, of course, existed before the construction of this neologism and was a feeling familiar to those living before 1688. What Hofer did was assess it as a disease that required treatment that could include medication, but was often assumed to be best addressed simply by returning the sufferer to their homeland. However, a return was not always as easy as making a journey, as the lost place was often temporally rather than spatially distant. This is certainly visible in early modern England. A pining for the common religious belief that existed before the schism of the Reformation has been traced in the writings of John Stow, for example—along with his nostalgic recollection of Goodman's dairy, of course.[3] Complaints about a "golden, vanished age of generosity" were regularly voiced in the period, too, and were often linked to changes in the world of agricultural production, with the breakdown of community seen as a product not only of urbanization but also of the buying up of the land by those, like Samuell Willcockes, perhaps, who lacked customary right.[4] Thus Shakespeare's late-sixteenth-century theater audience would have understood John of Gaunt's deathbed denunciation of his sovereign in *The Tragedy of King Richard the Second* as having contemporary resonance: England, Gaunt stated, "Is now leased out . . . / Like to a tenement or pelting [paltry] farm." He described Richard II's actions

---

2. Carolyn Kiser Anspach, "Introduction" to Hofer, "Medical Dissertation," p. 377. The score of the "Kühe-Reyen" (kuhreihen), is reproduced on p. 389. Jean-Jacques Rousseau included this in his *Dictionary of Music* (ca. 1778), where it is called the *ranz-des-vaches*. Commenting on the tune's ability to produce "bitter sorrow for the loss of" former pleasures in the hearer, he stated: "The music does not in this case act precisely as music but as a memorative sign." Quoted in Jean Starobinski, "The Idea of Nostalgia," translated by William S. Kemp, *Diogenes* 14 (1966): 92.

3. Susan Brigden, "Religion and Social Obligation in Early Sixteenth-Century London," *Past and Present* 103 (May 1984): 73.

4. Felicity Heal, "The Idea of Hospitality in Early Modern England," *Past and Present* 102 (February 1984): 80. See also Andy Wood, *The Memory of the People: Custom and Popular Senses of the Past in Early Modern England* (Cambridge: Cambridge University Press, 2013), p. 71; and David Graeber, "Manners, Deference and Private Property in Early Modern Europe," *Comparative Studies in Society and History* 39, no. 4 (1997): 723.

as undercutting the stability of past practices by opposing inheritance (possession) with commerce (leasing):

> O, had thy grandsire with a prophet's eye
> Seen how his son's son should destroy his sons,
> From forth thy reach he would have laid thy shame,
> Deposing thee before thou were possessed,
> Which art possessed now to depose thyself.
> Why, cousin, wert thou regent of the world
> It were a shame to let this land by lease. . . .
> Landlord of England art thou now, not king.[5]

Richard did not heed Gaunt's advice, but the audience would have recognized the language Shakespeare chose for the accusation. It might not be the monarch they regarded as the problematic landowner, however, but graziers and farmers: the former were those who earned a living fattening animals for sale rather than husbanding them from birth, the latter farmed on a large scale. Both were in it for the money and were understood not to be interested in working the land to maintain a household honestly, like laborers. Rather, as Robert Greene put it in *A Quip for an Vpstart Courtier*, graziers and farmers squeezed the countryside for capital and did not think of communities of people or of animals. He wrote of the grazier that he "wringeth leases of [pastures and meadows] out of poore mens handes," and of the farmer that by bidding "the Landelord farre more then the poore man paise yearlie for it," he took over tenancies, leaving those who were evicted to "begge in the street."[6] In 1594, two years after Greene's work was published, Will Kemp's *A Knacke to knowe a Knaue* noted something similar and invoked the figure of Piers Plowman to complain against the "vnknown Farmer" who is described as "a theefe that robs the common wealth."[7] His being unknown—a stranger

---

5. William Shakespeare, *The Tragedy of King Richard the Second*, in *William Shakespeare: The Complete Works*, edited by Stanley Wells and Gary Taylor (Oxford: Clarendon, 1988), 2.1.60–61,104–10, and 113. William O. Scott writes that "although Richard is still nominal owner of the realm, he has leased out the use of it to the tax farmers, who function as tenants (perhaps with the people as subtenants)." Scott, "Landholding, Leasing and Inheritance in *Richard II*," *SEL: Studies in English Literature 1500–1900* 42, no. 2 (2002): 276.

6. Robert Greene, *A Quip for an Vpstart Courtier* (London: Iohn Wolfe, 1592), G3v and G4r.

7. Will Kempe, *A Knacke to knowe a Knaue* (London: Richard Iones, 1594), C3r. The plowman was felt to be particularly impacted by increasing pastoral over arable farming during the period and was used to comment on agricultural change for this reason. See Andrew McRae, "Fashioning a Cultural Icon: The Ploughman in Renaissance Texts," *Parergon* 14, no. 1 (1996): 188–95.

to the soil—alongside his farming (profit-making), was presented as upsetting the stability of rural life with devastating implications.

Greene and Kemp were not alone: in the early seventeenth century, Nicholas Breton's Mad-Cappe likewise pined for a lost "golden time" during which "no man lou'd his neighbour to an end, / But once and euer, say and hold a friend." This ideal world of neighborliness was reckoned in terms of the fair use of real estate and communal care for quick cattle:

Then, was the *Sheepe* knowen easely by his *brand*,
Cow by her *lowe*, and by his *barke* the *Dogge*:
The *neighbour* iustly measur'de out his land,
And helpt to pull his *Horse* out of the *bogge*.

By the end of the poem, Mad-Cappe recognizes that he is now in a golden age of a different kind—"Such as haue *Golde*," he writes, "are in the *golden vaine*, / While that the *poor* must champe vpon their bit."[8] This is a world in which money counts for more than neighborliness; in which wealth separates the rich from the poor, who find themselves bridled like beasts: working the land but not owning it. It is also a world in which animals suffer because quick cattle are now for quick profit. For Mad-Cappe, the country ways of day-to-day, face-to-face engagements between rich and poor and between people and their animals are lost, and with it are extinguished ideas of community that cannot be replaced.

Given the apparent sense of danger that was regarded as attendant on the rise of these occupations in the countryside in this period, it may be unsurprising that there are no testators who term themselves graziers in the large ERO or LMA datasets and only four who list themselves as farmers—all from the large ERO dataset.[9] This small number of testators might signal either that the occupations were so loathed that almost no one would own up to them or that they were not, as yet, regular occupational terms in the area: after all, while there were four farmers, there were, by comparison, 1,121 testators who termed themselves yeomen in the county.[10] Only one of the

---

8. Nicholas Breton, *Olde mad-cappes new gally-mawfrey* (London: Richard Iohnes, 1602), C3r, E2r, and E2v.

9. It is perhaps notable that all four farmers were from parishes within the ERO jurisdiction of the Archdeaconry of Essex, all of which are within thirty-five miles of the capital.

10. One will raises the possibility that the terms yeoman and farmer (and thus the old ways and the new) were not so distinct as Greene's and Kemp's work might seem to imply. In his 1623 will, Thomas Pie of Rochford's occupation was listed first as yeoman but that was crossed through and replaced afterward in the same hand with the word farmer. See ERO, D/AEW 17/115 (1623), made probate 93 days after writing.

four farmers included specific animal bequests in his will, and the 1634 will of Silvester Pett of Mundon does seem to reflect some of the practices that are outlined in Greene's and Kemp's complaints about changes in the country-side. Pett, apparently childless,[11] bequeathed to his wife Anne "eleven milch Cowes, & fourtie Ewes & one Ramme being all now at the farme called Lang-meads North," and "my dunne horse." The numbers of cows and sheep are above the average for an Essex will, but otherwise this is familiar (indeed, he also left "one Russett Ewe lambe, wch I bought of John Pasfeild" to the daughter of a neighbor). It is in the detail of the real estate, however, that the will reveals Pett's status as farmer to have a potential meaning. In addition to being left the animals, a cock of hay, oats in the barn, and a bed and chair at Longmeads North, Anne also received "all my mooveable goods wthin my dwelling house at Iltney." This was not the collecting phrase, however. That phrase reads "All other my goods & Chattells, horse, kine, sheepe, Lambes, howsholdstuffe, mooveables upon the my two farmes ^the one^ called Jetner and the other Haisdons; And all other my possessions, ryghts, tytles and inter-estes to any farmes, lands, or leases, whatsoever, I giue & bequeath to Thomas Turner yeoman dwelling & scituate at ffeering in the Countie of Essex." In addition, in a codicil Pett requested that Turner plow and harrow a thirteen-acre piece of land at Langmeads North for his wife, "she finding him hay and chaffe for his horses during the time of the plowing & harrowing of the land." And, in the same codicil, he bequeathed "to Anne my wife one barrow hogge wch is at Iltney."[12]

Pett's will thus includes four properties (Longmeads North, Jetner, Hais-dons and Iltney) and perhaps more (it is not clear what "all other my posses-sions, ryghts, tytles and interestes to any farmes, lands, or leases, whatsoever" might refer to or whether they might be in different locations than those the will specifically mentions). And it is notable that while Iltney and Haisdons are both in Mundon,[13] the location of the other two properties is not clear. This means that either Pett (who lived in Mundon) or his legatee Turner

---

11. There is no surviving parish register from Mundon for this period to confirm this, but the will contains no bequests to immediate offspring.

12. ERO, D/AEW 19/299 (1634), made probate 50 days after writing. The other two farmers' wills are ERO, D/AEW 17/281 (1625), made probate 19 days after writing; and D/AEW 19/195 (1633), made probate 300 days after writing.

13. An early eighteenth-century "Memorandum of Surrender" in the Essex Record Office includes a reference to "Messuage and lands, customary and heriotable called Street House alias White House, Haysdons and Borehams (190 acres), formerly in occupation of William Brady and now of Thomas Goodson, copyhold of the manor of Mundon Hall in Mundon." ERO, D/DU 363/2 (1717), Essex Record Office, https://secureweb1.essexcc.gov.uk/SeaxPAM/ViewCatalogue.aspx?ID=957357, accessed 14 December 2016.

(who lived in Feering, eleven miles away) would be strangers on some of the land. If Jetner and Longmeads North were in Feering near Turner, then Pett would have been farming at a distance, but if they were in Mundon near Pett, then Turner would be doing that once he inherited the land. But the link between Pett's will and the complaints about changes to the countryside can be taken further. Iltney, his dwelling house, was named as the home of Thomas Gates, a husbandman, in Gates's 1623 will in which it was bequeathed to the oldest of his two young sons, John and Thomas. The document, specified that Thomas Gates senior's brother, James Forest, was to have and hold the "wholl lands & tenemts & other the estate whatsoeu" until both sons reached the age of 21.[14] The fact that the farm was owned eleven years later by Silvester Pett suggests that neither son survived to inherit the land and that their uncle sold it, perhaps directly to the farmer. The breakdown of inheritance, as John of Gaunt had feared, had taken possession outside of the family line and into commercial ownership.

Profiteering landlords and those who hoped to make money from agricultural production through ownership of distant property had, of course, been around for a long time,[15] and life in the countryside was never easy—for people or for animals. The working day Gervase Markham outlined in 1620 was idealizing, but it also revealed the sheer hard work that was required from the men and women, not to mention animals, involved.[16] And life was not always comfortable, either. Healthcare was rudimentary, and the weather might have had much greater impact than it does now. In 1607, a pamphlet reported that flooding had killed many animals and people in Somerset, Norfolk, Lincolnshire and Kent, and Dekker's North-Country-man bewailed the impact of the "great Snow" on his household-family seven years later.[17] It is against such realities that the nostalgically invoked ideal of the stability and generosity of the countryside by writers such as Breton must be placed.

The point of such looking backward, however, is not to list facts but to reconstruct a past that fulfills the needs of the present, and this is where nostalgia comes in. Its power, in Fred Davis's terms, is that it "attends . . . to the pleas for continuity and to the comforts of sameness": it offers a glimpse of

---

14. ERO, D/AEW 17/157 (1623), made probate 77 days after writing.

15. See, for example, Jeremy McInerney's discussion of the increasing commercialization of the farming of sacred cows in *The Cattle of the Sun: Cows and Culture in the World of the Ancient Greeks* (Princeton and Oxford: Princeton University Press, 2010), pp. 173–95.

16. Gervase Markham, *Markhams Farewell to Hvsbandry* (London: I. B. for Roger Jackson, 1620).

17. Anonymous, *1607. A true report of certaine wonderfull ouerlowings of Waters* (London: Edward White, 1607); Thomas Dekker, *The cold year. 1614. A deepe snow: in which men and cattell haue perished, to the generall losse of farmers, grasiers, husbandmen, and all sorts of people in the countrie; and no lesse hurtfull to citizens* (London: W. W., 1615), B2v.

how things can be imagined to have been in order to argue for how they should be once again.[18] And it is unsurprising that early modern writers, who were aware of their increasingly urban readerships, found a foundation from which to critique their world in the idealization of the countryside. The change from country ways was used as a marker of a general decline away from neighborliness, generosity, and simple living to self-centered ambition, profiteering, and the artificiality of city life. Indeed, it is perhaps ironic that in Nicholas Breton's urbane dialogue from 1618, *The Court and Country*, one of the ways country life was idealized was through the fact that there the written word was represented as being used, not for artifice or entertainment, but for the expression of faith and true record. The Countryman states:

> with vs, this is all we goe to schoole for: to read common Prayers at Church, and set downe common prises at Markets, write a Letter, and make a Bond, set downe the day of our Births, our Marriage day, and make our Wills when we are sicke, for the disposing of our goods when we are dead: these are the chiefe matters that we meddle with.

In response, Breton's Courtier, challenging the rural species boundary in his emphasis on his own courtly ways, asks: "What? is man but as a beast, bred like a fore-horse to goe alwayes right on, and rather draw in a cart, then trot in a better compasse?"[19]

Breton's text clearly comes out on the side of the Countryman for whom stability was not figured as stasis and stagnation but as fruitful. Indeed, the desired life was represented as one being lived alongside animals:

> We haue hay in the barne, horses in the stable, oxen in the stall, sheepe in the pen, hogges in the stie, corne in the garner [storehouse], cheese in the loft, milke in the dairy, creame in the pot, butter in the dish, ale in the tub, and *Aqua vitae* in the bottle, beefe in the brine, brawne in the sowce, and bacon in the roofe, hearbs in the garden, and water at our doores, whole cloths to our backes, and some money in our cophers, and hauing all this, if we serue God withall, what in Gods name can we desire to haue more?[20]

It was people's growing separation from animals, in short, that was used to exemplify all that was wrong with the world in the late sixteenth and early

---

18. Fred Davis, "Nostalgia, Identity and the Current Nostalgia Wave," *Journal of Popular Culture* 11, no. 2 (1977): 420.

19. Nicholas Breton, *The Court and Country, Or A Brief Discourse Dialogue-Wise Set Down Between a Courtier and a Countryman* (London: G. E. for John Wright, 1618), A4r, A4v, B3v, C2v, B3r–v.

20. Breton, *The Court and Country*, B2v.

seventeenth centuries, and the sound of the Kühe-Reyen was to become the medical marker of this.

But nostalgia was not only felt in the past: Bryan S. Turner has outlined what he calls the "nostalgia paradigm" of the modern world.[21] Such modern nostalgia is no longer viewed as a disease, of course—that was an idea that stopped being taken seriously by the end of the nineteenth century when, Jean Starobinski writes, "bacteriology and pathological anatomy" had come to underpin medical understanding and nostalgia gained a more "poetic" meaning.[22] But despite that shift, nostalgia remains a powerful feeling. For Turner there are four main aspects of it in the modern world. First is a sense of historical decline and a loss of "homefulness"; second, "a sense of the absence or loss of personal wholeness and moral certainty"; third, a feeling of "the loss of individual freedom and autonomy with the disappearance of genuine personal social relationships"; and fourth, "the idea of a loss of simplicity, personal authenticity and emotional spontaneity."[23] What comes with nostalgia, Turner and Georg Stauth argue, is "the myth of premodern stability and coherence."[24] Even without the late seventeenth-century sense of it as a disease, this modern version of nostalgia has been closely associated with animals. Hal S. Barron, for example, has argued that rural life was romanticized and associated with childhood in the silent films of early twentieth-century America at a time when an increasing proportion of the population was living outside the countryside (1920 was, according to census records, the first time in U.S. history that the majority of inhabitants lived in urban areas). Barron shows that over and over again, the cinema presented the pastoral world as "a lost Edenic past" of wholesome morality that contrasted with the dangerous nature of the city.[25]

The movement away from the rural environment to the city sits alongside the increasing industrialization of agriculture with, for example, the

---

21. Bryan S. Turner, "A Note on Nostalgia," *Theory, Culture and Society* 4 (1987): 150.

22. Starobinski, "The Idea of Nostalgia," pp. 100–101.

23. Turner, "A Note on Nostalgia," pp. 150–51.

24. Georg Stauth and Bryan S. Turner, "Nostalgia, Postmodernism and the Critique of Mass Culture," *Theory, Culture and Society* 5 (1988): 512.

25. Hal S. Barron, "Rural America on the Silent Screen," *Agricultural History* 80, no. 4 (2006): 384, 395. Buster Keaton's 1925 film *Go West* problematizes this narrative by recognizing the commercialization of ranching that has taken place. The film tells the story of Homer Holiday (aka "Friendless," a fitting name for a man who has migrated west to make his fortune) and his attachment to Brown Eyes, the film's only named cow. This attachment sets Friendless apart from the other ranch hands. At the film's climax, Friendless saves Brown Eyes from the slaughterhouse, leaving all the other (unnamed) cattle to rampage through the streets of Los Angeles, thus symbolically challenging the nostalgic vision that regarded rural life as distinct from urban living that so many other films contained. *Go West*, dir. Buster Keaton, 1925, distributed by Metro-Goldwin-Meyer.

mechanization of milking that I have proposed had its (human) beginnings in the commercialization of the role of the milkmaid in the seventeenth century fully (mechanically) under way by the end of the nineteenth century.[26] Shifts such as these distanced people from animals and at the same time, it seems, created a new feeling for the past, and the early modern (preindustrial, pre-mechanized) smallholding was regarded as a place of—to use Turner's terms—"homefulness," moral certainty, genuine personal relationships, and personal authenticity. In short, preindustrial farming came to be nostalgically considered a site of mutually satisfying order at the moment when that order was felt to have been lost.

It is in this context that the concept of the social contract might usefully be read. I have shown that much that Bernard E. Rollin suggests about this holds true for thinking about early modern people's relationships with their animals. Indeed, when he outlines the new ethic that he hopes will displace the object status that animals hold in industrial agriculture, Rollin calls up a vision that would be recognizable in meaning, if not terminology, to many early modern people. This new ethic, he writes:

> is conservative, not radical, harking back to animal use that necessitated and thus entailed respect for the animals' natures. It is based on the insight that what we do to animals *matters* to them, just as what we do to humans matters to them, and that consequently we should respect that mattering in our treatment and use of animals as we do in our treatment and use of humans.[27]

This is calling us to acknowledge a world of affinities, and the appeal to return to an earlier way of farming emphasizes the lost closeness of humans and animals in order to respond to current practices that, for example, assume that the working life of a cow might only last for "slightly longer than two lactations [because of] metabolic burnout and the quest for ever-increasingly productive animals" and that still, in some places, keep lactating sows in crates so small that they cannot move.[28] What matters to these animals has clearly

---

26. See G. E. Fissell, "The Evolution of Farm Dairy Machinery in England," *Agricultural History* 37, no. 4 (1963): 217–24.

27. Bernard E. Rollin, *Farm Animal Welfare: Social, Bioethical, and Research Issues* (Ames: Iowa State University Press, 1995), p. 18.

28. Bernard Rollin, 'The Ethics of Agriculture: The End of True Husbandry,' in *The Future of Animal Farming: Renewing the Ancient Contract*, edited by Marian Stamp Dawkins and Roland Bonney (Oxford: Blackwell, 2008) p. 13. Joyce D'Silva describes the life of the modern dairy cow: "A couple of months after her first calf, she's pregnant again and continues to be milked until a few weeks before giving birth again, thus spending around eight months a year both pregnant and lactating. Frantic for food to sustain this level of output, the cow requires concentrated rations to supplement her diet, yet

been evacuated in the name of productivity, with commerce overriding not the inheritance of property, but the inheritance of (human and animal) ways of living.

Rollin's idea about going back to a way of collaborating with animals that recognizes that what matters to them should matter to us reflects the sense I have traced in early modern writings that husbandmen and house-wives understood that the needs and desires of their animals had to matter for successful production. But Rollin's reading is also what we might call a kind of hortatory history. It calls on us to think about the past in order to return to it because it claims that it offers a better ethical (although not com-mercial) way of producing meat. Its sense of that past's prioritizing of care over money, of affinity over profit, however, hints at a problem: it is remi-niscent of the late-sixteenth-century complaints about graziers and farmers and reveals, perhaps, that the ideal human-animal encounter in agriculture was and remains just beyond the horizon, in an imagined past that is forever out of reach.[29]

The work of the animal welfare scientist Françoise Wemelsfelder offers a rather different way of thinking about the relationship between past and present practices that is constative rather than hortative, but what it reports makes problematic in a different way any call for a return to the method that underpins the historic social contract. Wemelsfelder's Qualitative Behavioural Assessment methodology (QBA) calls on farm assessors (that is its practical application) to look at animals as what she calls "whole creatures." Thus, instead of transforming an animal's "experience into a 'thing,' an object for study,"[30] the aim of QBA is to ask observers to

> perceive not merely a string of separate "behaviours," but the unity of those behaviours . . . which is the "behaver," the animal. This "behaver" does not just emerge from the sum total of observed separate behav-iours; it executes these behaviours in a certain manner, and it is this

these same rations can be a factor in the appallingly high levels of lameness seen in cows . . . with 25% of cows requiring treatment for lameness every year. But not all cows get treated when lame." D'Silva, "The Urgency of Change: A View from a Campaigning Organization," in Dawkins and Bonney, ed., *The Future of Animal Farming*, p. 36.

29. This is reminiscent of the sense that the cannibals were never actually encountered but only ever heard of; that they were always living on the next island. See Anthony Pagden, *The Fall of Natural Man: The American Indian and the Origins of Comparative Ethnography* (Cambridge: Cambridge University Press, 1982), pp. 82–83.

30. Françoise Wemelsfelder, "A Science of Friendly Pigs . . . Carving out a Conceptual Space for Addressing Animals as Sentient Beings," in *Crossing Boundaries: Investigating Human-Animal Relation-ships*, edited by Lynda Birke and Jo Hockenhull (Leiden and Boston: Brill, 2012), p. 225.

instrumental [i.e., active] relationship that gives the animal's movement its expressive character.[31]

Wemelsfelder is challenging assumptions about animal behavior made by scientists who do not view the animal's experience as an extended, cumulative thing but turn it instead into a "string of separate 'behaviours.'" Eileen Crist writes that such scientists "de-sequence" animals' worlds; they do not regard animals' actions as evidence of an experiencing being's volition, but as "mere movement." Crist shows how behaviorist scientists view honeybees, for example, as "exist[ing] in a perennial moment, propelled by impinging stimuli to move through disjointed pockets of space." Such a reading proposes that they "do not *use* odor as a guide, rather the *odor* guides them to the food."[32] For behaviorists, bees are what Crist calls "natural objects"—they are understood to lack reason and instead are regarded as "compulsively steered by (interior-physiological and/or exterior-environmental) stimuli beyond their control or comprehension."[33] From such a perspective, a bee cannot be understood to be actively scenting out pollen; instead, it is driven toward the pollen by irrepressible forces, and one of the possible implications of this is that such a creature can never be comprehended as having the capacity to scent out and then reject some odors, such as the smell of garlic or stale alcohol or sweat. The behaviorist perspective does not allow for that; it denies animals a sense of their own agency in the construction of their experiences.

This is an extreme offshoot of the mechanistic perspective that is the default position of much current animal science and that has its foundation in the mid-seventeenth-century work of René Descartes. This perspective, in Wemelsfelder's words, "regards automated, rule-based causation between physical elements as the main organizing principle through which living beings emerge as functional biological systems"; it assumes, as already noted, "that animals can evolve to high levels of complexity without needing subjective perspective or experience, unless evidence is found to the contrary."[34] In a challenge to the mechanistic view, however, Wemelsfelder and her colleagues propose not only that animals drive their own activities but that they experience them too, in ways that are, once again, affined to

---

31. Françoise Wemelsfelder, Tony E. A. Hunter, Michael T. Mendl, and Alistair B. Lawrence, "Assessing the 'Whole Animal': A Free Choice Profiling Approach," *Animal Behaviour* 62, no. 2 (2001): 218–19.

32. Eileen Crist, "Can an Insect Speak? The Case of the Honeybee Dance Language," *Social Studies of Science* 34, no. 1 (2004): 32.

33. Eileen Crist, *Images of Animals: Anthropomorphism and Animal Mind* (Philadelphia: Temple University Press, 1999), pp. 6, 9.

34. Wemelsfelder, "A Science of Friendly Pigs," p. 225.

the ways humans experience their activities. They write: "Human observers can integrate perceived behavioural details and signals to judge an animal's behavioural expression, using qualitative descriptors (e.g. relaxed, anxious) that reflect the animal's affective (emotional) state."[35] That is, in observing the whole animal, QBA allows what might seem to be anthropomorphic projections of human experience onto animals ("that pig seems confident," "that cow is nervous") to be read as in fact repeatable and valid assessments of animal behavior: repeatable because different people have been shown to make similar evaluations; valid because there are what Wemelsfelder has termed "strong and complex correlations between QBA and other accepted scientific measures of the animal's behaviour and physiology."[36] In short, QBA reveals accusations of anthropomorphism to be unfounded. Indeed, it refutes the idea of anthropomorphism altogether in that it claims that attributing to animals feelings of happiness, boredom, fear, and so on is not projecting human emotions onto them but is, in fact, identifying states that the animals themselves experience. QBA recognizes, for example, that bovine anxiety exists in order to work to address it. To reject such a description as simply anthropomorphic would be to regard that anxiety as a mere projection of an unscientific human mind and would be, in turn, to dismiss a cow's disquiet as not real and so not worth addressing.

The ethical implications of the shift from the behaviorist to the QBA perspective are significant. Instead of de-sequencing animals, QBA calls on observers to recognize them as agents who live in sequential time—capable of remembering, of responding to the world, to each other and to the other beings in it—and able to change their world to reflect their own desires. It also asks observers to assess animals' welfare in relation to this recognition. We have encountered such animals and such assessors in early seventeenth-century Essex. The cow who refused to be milked unless she was sung to is a cow who recalled being sung to, remembered a pleasure in that experience, and wished it to be repeated. The maid who responded with a song

---

35. Kenneth M. D. Rutherford, Ramona D. Donald, Alistair B. Lawrence, and Françoise Wemelsfelder, "Qualitative Behaviour Assessment of Emotionality in Pigs," *Applied Animal Behaviour Science* 139, nos. 3–4 (2012): 218.

36. Wemelsfelder, Hunter, Mendl and Lawrence, "Assessing the "Whole Animal." Further validation of the method is outlined in F. Wemelsfelder and S. Mullan, "Applying Ethological and Health Indicators to Practical Welfare Assessment," *OIE Scientific and Technical Review* 33, no. 1 (2014): 111–20. One QBA study described how pigs that had been injected with Azaperone, a drug used to prevent aggression and stress, were identified by observers as being more "confident/curious" than those who had been injected with a saline solution. The descriptions of the observers correlated closely with the physiological data despite the fact that the observers were not aware that the injections had been given. Rutherford, Donald, Lawrence, and Wemelsfelder, "Emotionality in Pigs."

assessed the cow as an agent capable of feeling emotions in her (the maid's) world and adjusted that world to fit that recognition. Likewise, the injunctions about cleanliness when approaching bees were advising readers based on an acknowledgment that bees had their own experiences and could and would express them, and that beekeepers needed to understand those expressions to be successful.

Instead of just being regarded as an important modern conception of animal behavior, then, QBA might also be seen as a theory of animal being that is more in line with the beliefs of the early modern men and women who have been the focus of this book than any mechanistic view might be. That is not an anachronistic proposal; it is actually historically appropriate because QBA marks a shift away from the assumptions that have underpinned industrial farming methods since the mechanistic perspective achieved dominance. That perspective, as noted, was being born in the seventeenth century. Even as allegory (with its potential for the expression of active animal agency) was being removed to the realm of the literary by the followers of a father of empirical science, Sir Francis Bacon,[37] so evaluation of the subjective experience of animals was being repressed by the followers of a father of modern philosophy, René Descartes; and it is these intellectual developments that produced behaviorist ideas about animals. Such shifts had not found their way into the yards and fields of the testators in the large ERO dataset, had not been naturalized in parts of early modern Middlesex in the way that they have now. Instead, the beekeeping texts from that period emphasized analogies and similes and traced apian morality; and the descriptions of milking focused on the collaborative presence of the cow. Thus, QBA (albeit without the apian moralizing) represents a kind of return to a way of viewing animals that has been ignored by centuries of orthodox mechanistic animal science.

Vinciane Despret offers another approach that can be used to contemplate the relationship between QBA and past agricultural practices. This also calls for a return to a lost way of thinking, but the sense of return in Despret's work is not, like Rollin's, nostalgic—it is not a backward projection, an attempt to retrieve "homefulness," moral certainty, genuine personal relationships, and personal authenticity in an imagined past. Instead, Despret more actively recognizes that the histories that construct, and you might say constrict, our thinking play a role in what we are able to contemplate now.[38] She looks to

---

37. See Peter Harrison, *The Bible, Protestantism, and the Rise of Natural Science* (Cambridge: Cambridge University Press, 1998), p. 91.

38. As Donna Haraway has written, "It matters what matters we use to think other matters with; it matters what stories we tell to tell other stories with; . . . what thoughts think thoughts." Haraway,

the ideas of Charles Darwin—who wrote of the "real affinities of all organic beings,"[39] and who linked humans and animals in terms of their emotions as well as their bodies—and finds in his work the possibility of "stories of connivances, attractions, reciprocal inductions, and also repulsions, that weave their own narratives in the web, and therefore create new connections and affinities." Such narratives, she finds, are now being used to challenge the dominance of mechanistic thinking in the ecological sciences.[40] Despret does not call for "enchantment, nor [for] re-enchantment [of the world]. This would imply," she argues, "that the world originally would have been non-enchanted, or previously disenchanted." What we need to do instead, she says, is learn to think *"before the disenchantment happens."*[41] That is, we need to acknowledge that our construction of the world emerges from a rejection of a certain worldview and that that rejection should be recognized as historically constituted and not natural. This would allow us to recognize that allegory has not always been found in the realm of the literary, for example; that animals have not always engaged only mechanically in work.

From this perspective, QBA can be read as existing in a relationship with the past, not, as with Rollin's work, in a knowing way that deliberately invokes an idealized bygone age for ethical purposes. Instead, Wemelsfelder and her colleagues are undertaking scientific research (as that is understood today), and in the process are recovering a sense of animals that has been marginalized by mechanistic thinking. They are finding creatures with whom one might converse and collaborate, to whom one might sing. Such "recoveries" will come as no revelation to people who live with and work closely with animals, of course—their sense of the collaborative relationship between themselves and their animals is frequently voiced, as Manuel Calado Varela knew when he opened the door for his cows. What this signals is the separation between lived relation and politics, between hands-on, day-to-day experiences and what can stand as proof in the construction of laws. In such a context, Rollin's call for a transformation in human-livestock relations using the legal

---

Staying with the Trouble: Making Kin in the Chthulucene (Durham, NC, and London: Duke University Press, 2016), p. 12.

39. Charles Darwin, *On the Origin of Species*, edited by Gillian Beer (Oxford: Oxford University Press, 2008), p. 352.

40. Vinciane Despret, "From Secret Agents to Interagency," *History and Theory* 52, no. 4 (2013): 35. She is referring to the work of historians of science Carla Hustak and Natasha Myers.

41. Despret, "From Secret Agents to Interagency," p. 36. Despret highlights the importance of Darwin's work, but I am suggesting that the fields and yards of early modern Essex might offer earlier evidence for such thinking and that perhaps Darwin was also engaged in an act of reading "before the disenchantment happens" when he claimed affinities among species and was—like the QBA scientists after him—giving scientific validity to what had been known back then by experience.

concept of the contract seems even less likely to succeed. What is perhaps crucial is that, beyond the recognition that animals live in sequential time, that they can remember, respond, and affect changes that reflect their own desires, Wemelsfelder and her colleagues are proving this in terms that can be accepted by the wider scientific community, and it is the scientific community that holds the key to changes in welfare, so dominant has the mechanistic perspective become. Anecdotal evidence—the sows don't like this; the cows get upset if we do this—is not enough to instigate changes to welfare standards. Scientifically valid proof is required nowadays to establish the presence of animals as collaborators, and QBA is beginning to offer that.

But there is still a problem: indeed, the problem is likely to become more pressing because of the findings of QBA. In chapter 4, I highlighted a limit to the social contract that was encountered daily in the slaughter of animals. Rollin's conception, like so many visions of the past, pushes us to think about former times as an era of another way of being in which homefulness, moral certainty, genuine personal relationships, and personal authenticity were present. But in offering this reading of the past, Rollin omits to mention the aspects of that age that have been lost between then and now, that might make the temporal distance between the two impossible to bridge. I am not voicing this as a problem of his reading of the past (I am not criticizing Rollin as a poor historian), but as a problem for his outlining of the means by which the revision of human-animal relations might be brought about. And it is a problem, I would suggest, that represents an impasse in current thinking about our relationships with animals in agriculture.

The industrialization of agriculture has removed some of the possibilities for close encounters between people and animals and has shortened animals' lives, for sure; but these shifts can be addressed (even if animals' bodies, so altered by breeding practices, can never be returned to an earlier state).[42] The changes that would allow for closer individual attention to and longer lives for animals would undoubtedly impact the profit margin of agricultural production and the kind of food that might be produced, but the changes could happen if there was a desire to do it. What will be more difficult to attend to, however, is the disappearance of the emblematic worldview that presented animals as having more than material meaning: that, you might say, recognized them as enchanted beings. Without the potential to view animals as possessing more than worldly meaning

---

42. One of the problems of shifting battery chickens to free-range farms is that the breed of chicken that is good in one production unit might not be good in another. See Mara Miele, "The Taste of Happiness: Free-Range Chicken," *Environment and Planning A* 43 (2011): 2076–90.

(a central plank of preindustrial agriculture), it is difficult to see how the mode of husbandry with which that view coexisted—one of close collaborative relationships between small numbers of humans and small numbers of animals—can be reclaimed, because one requires the other. And our growing understanding of animals' capacities—for friendship, culture, pleasure, boredom, and so on—has made this problem more pressing.

In ancient Greek culture, as Jeremy McInerney suggested, sacrifice was a means by which the paradox of pastoralism was addressed.[43] Giving slaughter a sacred role deflected the betrayal that killing agricultural animals represented and gave significance to the animal, and to the act that might otherwise have undermined the meaning of husbandry. I have proposed that in early modern England the emblematic worldview had a similar function: reading animals as meaningful in both worldly and otherworldly terms made their slaughter part of their larger role. Both of these means of addressing the problem of killing looked outside animals to find meaning—to the realm of the supernatural, to religion. The worldview that QBA presents, in contrast, has placed their meaning within the animals themselves. There is no need to invoke the supernatural here; rather, it is animals' consciousness, agency, feelings—their natures—that are important, not their status as exemplars or symbols of an external divine authority. And in this context, killing them can be regarded only as taking from them—violently or otherwise—that which makes them who they are. Slaughter is no longer a fulfilment of a divine narrative; it is not the completion of the animals' inseparably worldly and otherworldly role. Killing is the curtailing of the life of an experiencing being who has skills, competences, and desires.

Studies of current agriculture reveal the absence of this worldview and the impact of that absence. They show that people directly involved with animals have not simply adopted the mechanistic perspective that replaces animal feeling with an assumption of "automated, rule-based causation between physical elements," even though that perspective might remove the horror of slaughter—as Descartes wrote in defense of his hypothesis that animals were natural automata in 1649, "my opinion is not so much cruel to animals as indulgent to human beings . . . since it absolves them from the suspicion of crime when they eat or kill animals."[44] Despite the theoretical dominance of the perspective that emerged from Cartesian thinking, farmers and

43. McInerney, *The Cattle of the Sun*, p. 36.

44. René Descartes to Henry More, February 1649, in *The Philosophical Writings of René Descartes*, edited by John Cottingham, Robert Stoothoff, and Dugald Murdoch (Cambridge: Cambridge University Press, 1985), 3:366.

slaughterhouse workers, like their early modern forebears, often engage with animals as co-workers rather than as objects, and the discomfort caused by the need to "recommodify" an animal—to shift it conceptually from being an individual with whom one has emotional relations to a product that is saleable[45]—can only be enhanced by the growing scientific understanding of and welfare attention being paid to animal feelings.[46] Rhoda Wilkie's study of stockmen and women in twenty-first-century Scotland, for example reveals that while the scale might be different in modern farms, some things have not changed from early modern times. Individualization and collaboration are still central to daily life with animals because the animals remain responding beings, even as they might now be numbered rather than named. For this reason, Wilkie shows, distancing maneuvers are sought by the people who work with them at the moment livestock begin their journey to becoming deadstock.[47] Likewise, Mark Riley, who has looked at dairy farmers' experience of retirement and the impact it has on their animals, has noted that "the subjectifying/objectifying boundary between farmers and livestock is a diffuse, and arguably permeable, one." He recognizes that preparing animals for sale is "not simply material—that is preparation through grooming or fattening of animals—but also discursive," requiring the production of "one or more strategies . . . to allow a moral and empirical separation from the animal as it became positioned as a commodity."[48] Ensuring that the connecting word in the sentence "A peasant becomes fond of his pig and is glad to salt away its pork" remains an "and" and not a "but," as John Berger put it, does not appear to be straightforward.[49]

It is easy to understand, when seen like this, how those who want to improve the lot of animals in agricultural production might regard a world in which that difficulty seemed to be absent as ideal. A restoration of such a world would be a return to a way of living in which farming might be understood as being based on a contract in which animals were recognized as individuals who were understood to be capable of understanding and treated

45. Mark Riley, "'Letting them Go'—Agricultural Retirement and Human-Livestock Relations," *Geoforum* 42, no. 1 (2011): 16.

46. Mara Miele notes that "even if the 'happy chicken' might be seen as a weak invention, fraught with ambivalence and ambiguities, it has achieved important effects: it has suggested a more complex moral relationship between human and nonhuman animals, a search for different intimacies between humans and chickens." Miele, "The Taste of Happiness," p. 2087. Welfare assessments of happy chickens include the use of QBA methodology (p. 2086).

47. Rhoda Wilkie, *Livestock/Deadstock: Working with Farm Animals from Birth to Slaughter* (Philadelphia: Temple University Press, 2010). On distancing, see pp. 129–46.

48. Riley, "'Letting them Go,'" p. 20. For many farmers Riley interviewed, retirement was a disruption of the rules of farming and thus made the recommodification of animals more difficult and created "moral unease" (p. 21).

49. John Berger, *About Looking* (London: Writers and Readers Publishing Co-operative, 1980), p. 5.

as such, in which eating meat did not require that animals be made invisible at the meal table, as Carol J. Adams has argued that modern consumption does,[50] but had importance because the act was acknowledged to be the consumption of an animal and it took responsibility for and gave meaning to that consumption. And it is perhaps unsurprising, given the recognition of animals' status as agents, as creatures whose experiences matter to them, that the way forward Rollin proposes is formulated using the language of the law, not because it objectifies animals (turns them into absolute property) but because it makes them parties to the contract and thus invokes a sense of the participation of two willing and active partners (one protecting and feeding, the other producing). Through this lens, the social contract seems to be a document of clarity that tidies up the problem of animal husbandry by calling up animals' presence as active beings without recourse to the supernatural.

But as the testators of early seventeenth-century Essex knew, things (cattle, chattels, goods) were never as straightforward as that. They saw their animals as affined, part of the household-family; and they understood them to make meaning in the spiritual and the material world, as Edward Osborne, a yeoman from Stow Maries, revealed accidentally when he bequeathed to his namesake son "one bible, one cow, one desk," in that order.[51] What this study of early modern ideas has shown is that a return to this lost world, to a time before the disenchantment happened, as Despret put it, is impossible because it would require more than legally mandated changes to animal housing and care, something more complex than simply turning back the clock.

It is perhaps strange that it is wills that can be read as offering this insight. A will, after all, would seem to tidy the world into controllable portions of stuff and illustrate the sense of animals as simply self-moving movable goods. What the wills of the early seventeenth-century testators in Essex actually bring to light is far from this, however. They reveal not a world of clear agreement (I feed, you produce) but a world of ongoing negotiation and tentative collaboration, a world, indeed, that was alive with individual wills (both human and animal) that were utterly entangled with each other; that was messy and uncomfortable, full of affection and instrumentality, life and death. This was a world in which a wife might have had a calf, a lamb might teach you about salvation, bees might have a moral code, and a cow might answer back. It is important, I think, to acknowledge that such a way of being was once possible, even if it has now, I suspect, departed permanently.

---

50. Carol J. Adams, *The Sexual Politics of Meat: A Feminist-Vegetarian Critical Theory* (Cambridge: Polity, 1990), p. 40.

51. ERO, D/AEW 19/227 (1633), made probate 4,429 days after writing.

❧ Bibliography of Primary Sources

Manuscript Sources

Essex Record Office (ERO)

Wills

Commissary of the Bishop of London (1620–1634)
    D/ABW 43/1–244
    D/ABW 44/1–348
    D/ABW 45/1–184
    D/ABW 46/1–249
    D/ABW 47/1–298
    D/ABW 48/1–261
    D/ABW 49/1–376
    D/ABW 50/1–262
    D/ABW 51/1–341
    D/ABW 52/1–225
    D/ABW 53/307

Archdeaconry of Colchester (1620–1634)
    D/ACW 8/240–305
    D/ACW 9/1–261
    D/ACW 10/1–278
    D/ACW 11/1–317
    D/ACW 12/1–119
    D/ACW 13/1–43

Archdeaconry of Essex (1620–1634)
    D/AEW 16/261–306
    D/AEW 17/1–329
    D/AEW 18/1–231
    D/AEW 19/1–299

Archdeaconry of Middlesex (Essex and Hertfordshire Jurisdiction) (1620–1634)
    D/AMW 1/23–277
    D/AMW 2/1–299
    D/AMW 3/1–250
    D/AMW 4/1–234

Archbishop of Canterbury: Peculiar of Deanery of Bocking (1620–1634)
    D/APbW 1/5–32

Bishop of London: Peculiar of Good Easter (1620–1634)
D/Apg W1/8–17

Parish Registers

D/P 59/1/1, Parish Register, St Mary the Virgin, West Bergholt (1559–1658)
D/P 60/1/1, Parish Register, St Peter, Roydon (1567–1706)
D/P 101/1/1, Parish Register, St Michael, Woodham Walter (1559–1789)
D/P 128/1/2, Parish Register, St Peter, South Weald (1559–1654)
D/P 135/1/2, Parish Register, Holy Trinity, Heydon (1558–1755)
D/P 140/1/1A, Parish Register, St Margaret, Stanford Rivers (1558–1745)
D/P 164/1/1, Parish Register, St Margaret, Tilbury-Juxta-Clare (1560–1724)
D/P 172/1/1, Parish Register, St Mary, Wix (1560–1675)
D/P 180/1/1, Parish Register, St Mary the Virgin, Little Easton (1559–1783)
D/P 210/1/1, Parish Register, St Swithin, Great Chishall (1583–1747)
D/P 230/1/1, Parish Register, All Saints and St Faith, Childerditch (1537–1710)
D/P 290/1/1, Parish Register, St Mary the Virgin, Farnham (1559–1635)
D/P 299/1/3, Parish Register, All Saints, Terling (1538–1688)
D/P 343/1/1, Parish Register, St Mary the Virgin, Mistley (1559–1696)
D/P 379/1/1, Parish Register, St Stephen, Cold Norton (1539–1771)
D/P 411/1/1, Parish Register, St Mary the Virgin, Matching (1558–1746)

Other

D/DBy A382, "Very Detailed Weekly Provision Book," Audley End Estate, Saffron
    Walden (1617–1629)
D/DMa M18, "A survey taken in September last 1621of the lands of Sr John Denham
    knight" (1621)
D/DP A52, "Thorndon Hall Accounts—Petre Family of Ingatestone" (1633–1642)
D/DP A74, "Household expenditure accounts, probably kept by a member of branch
    of the Petre family [of Ingatestone]" (1631–1640)
D/Dra Z1, Charles Chadwick's Common Place Book (1611–1627)

London Metropolitan Archive (LMA)

Wills

Diocese of London (1620–1634)
        DL/C/B/008/MS09172/031 98–318
        DL/C/B/008/MS09172/032 1–290
        DL/C/B/008/MS09172/033 1–216
        DL/C/B/008/MS09172/034 1–267
        DL/C/B/008/MS09172/035 1–227
        DL/C/B/008/MS09172/036 1–195
        DL/C/B/008/MS09172/037 1–269
        DL/C/B/008/MS09172/038 1–317
        DL/C/B/008/MS09172/039 1–327

DL/C/B/008/MS09172/040 1–315
DL/C/B/008/MS09172/041 1–134
DL/C/B/008/MS09172/042 1–166

OTHER

CLA/008/EM/02/01/001, "City Land Grant Book" (1589–1616)
DL/C 233, Diocese of London Consistory Court, Depositions Book (1630–1634)
P69/DUN2/A/002/MS10343, Parish Register, St Dunstan in the West (1558–1622)
P69/DUN2/B/030/MS03781, "Book of Depositions by inhabitants of the parish and
        others in the course of a law suit (Johson vs Wright), relating to tithes and
        tenths issuing from the Black Mule in Fleet Street" (1624)
P69/DUN2/B/061/MS03795, "Bonds of Indemnication and Releases" (1614)

National Archive (NA)

WILLS

Prerogative Court of Canterbury
        PROB 11/89/282 (1597)
        All Essex, London, and Middlesex wills in the range PROB 11/151/415–776 and
        PROB 11/152/10–866 (1627)

OTHER

STAC 8/34/5, "Attorney General v Nightingale" (1620)
WARD 2/55A/191/2, "Receipt from William Bindder and Edward Hall, to John
        Danyell of Hackney, Middlesex" (1600)

Printed Sources

*The Boke of common prayer and administration of the Sacramentes and other rites and
        Ceremonies in the Church of Englande.* London: Edwardi Whytchurche, 1552.
Boorde, Andrew. *The first and best part of Scoggins iests.* London: Francis Williams, 1626.
Brathwaite, Richard. *Whimzies: Or, a new cast of characters.* London: Felix Kingston,
        1631.
Breton, Nicholas. *The Court and Country, Or A Brief Discourse Dialogue-Wise Set Down
        Between a Courtier and a Countryman.* London: G. E. for John Wright, 1618.
———. *Old Mad-cappes new Gally-mawfrey.* London: Richard Iohnes, 1602.
Butler, Charles. *The Feminine Monarchy or Treatise Concerning Bees.* Oxford: Joseph
        Barnes, 1609.
Camden, William. *Britain, or A chorographicall description of the most flourishing king-
        domes, England, Scotland and Ireland, translated by Philemon Holland.* London:
        Iohn Norton, 1610.
Campion, Thomas. *Two Bookes of Ayres.* London: Tho. Snodham, 1613[?].
Chettle, Henry. *Kind-Harts Dreame.* London: William Wright, 1593.

Cicero. *The Booke of freendeship of Marcus Tullie Cicero.* Translated by John Harington. London: Tho. Berthelette, 1562.

Cockaine, Thomas. *A Short Treatise of Hunting.* London: Thomas Orwin, 1591.

Dalton, Michael. *The Covntrey Ivstice, Conteyning the practise of Ivstices of the Peace out of their Sessions.* London: Society of Stationers, 1618.

Dekker, Thomas. *The cold year. 1614. A deepe snow: in which men and cattell haue perished, to the generall losse of farmers, grasiers, husbandmen, and all sorts of people in the countrie; and no lesse hurtfull to citizens.* London: W. W., 1615.

Dent, Arthur. *The plaine man's path-way to heauen.* London: Melchiside Bradwood, 1607.

Descartes, René. *The Philosophical Writings of René Descartes.* Vol. 3. Edited by John Cottingham, Robert Stoothoff, and Dugald Murdoch. Cambridge: Cambridge University Press, 1985.

Dowe, Bartholomew. *A dairie Booke for good huswiues.* London: Thomas Hacket, 1588.

Estienne, Charles. *Maison Rustique Or, The Countrey Farme.* London: Adam Islip, 1616.

Evans, Sir William David. *A Collection of Statutes Connected with the General Administration of the Law.* London: Blenkarn, Lumley and Bond, 1836.

Fuller, Thomas. *The Historie of the Holy Warre.* Cambridge: John Witham, 1639.

Greene, Robert. *A Quip for An Upstart Courtier: Or, A quaint dialogue between Velvet breeches and Clothbreeches.* London: Iohn Wolfe, 1592.

Hall, Joseph. *Meditations and Vowes Diuine and Morall.* London: Humfrey Lownes, 1607.

Harington, John. *The Most Elegant and Witty Epigrams of Sir John Harington, Knight* London: G. P. for Iohn Budge, 1618.

Herbert, George. *A Priest to the Temple, or, The Countrey Parson his Character, and Rule of Holy Life.* London: T. Maxey, 1652.

Heresbach, Conrad. *Foure bookes of husbandry . . . Newely Englished, and Increased, by Barnabe Googe.* London: Richard Watkins, 1577.

——. *The Whole Art and Trade of Hvsbandry Contained In foure Bookes.* London: T. S., 1614.

Hill, Thomas. *A Profitable Instruction of the Perfect Ordering of Bees.* London: Edward Alde, 1593.

Jackson, Thomas. *Iustifying faith, or The faith by which the just to liue.* London: Iohn Beale, 1615.

Jonson, Ben. "On the Famous Voyage." In *Ben Jonson: The Complete Poems.* Edited by George Parfitt. London: Penguin, 1988.

Josselin, Ralph. *The Diary of Ralph Josselin, 1616–1683.* Edited by Alan Macfarlane. Oxford University Press, 1976.

Kempe, Will. *A Knacke to knowe a Knaue.* London: Richard Iones, 1594.

Lambarde, William. *Eirenarcha: or the office of the iustices of peace in foure bookes.* London: Ralph Newbery, 1592.

Lawson, William. *A new orchard and garden.* London: Bar. Alsop, 1618.

Levett, John. *The Ordering of Bees: Or, The Trve History of Managing Them.* London: Thomas Harper, 1634.

Lodge, Thomas, and Robert Greene. *A Looking Glass for London and England.* London: Barnard Alsop, 1617.

Lupton, Donald. *London and the countrey carbonadoed and quartred into seuerall characters.* London: Nicholas Okes, 1632.

Markham, Gervase. *Cavelarice, Or the English Horseman*. London: Edward Allde, 1607.

——. *Cheape and Good Hvsbandry: For the well-Ordering of all Beasts, and Fowles, and for the generall Cure of their Diseases*. London: T. S. for Roger Jackson, 1616.

——. *The English Hovse-Wife, Containing the inward and outward Vertues which ought to be in a compleate Woman*. London: Anne Griffin, 1637.

——. *Markhams Farewell to Hvsbandry*. London: I. B. for Roger Jackson, 1620.

Mascall, Leonard. *The Husbandlye ordring and Gouernmente of Poultrie*. London: Thomas Purfoote, 1581.

——. *The Gouernment of Cattell*. London: Tho. Purfoot, 1627.

Meres, Francis. *Wits common wealth The second part*. London: William Stansby, 1634.

Moore, John. *A Mappe of Mans Mortalitie*. London: George Edwards, 1617.

More, Thomas. *Utopia*. Edited by George M. Logan and Robert M. Adams. Cambridge: Cambridge University Press, 1989.

Norden, John. *Speculi Britanniae Pars: An Historical and Chorographical Description of the County of Essex*. 1594; repr., London: Camden Society, 1840.

——. *Speculum Britainniae. The first parte*. London: Eliot's Court, 1593.

Noy, William. *A Treatise of the Principal Grounds and Maximes of the Lawes of this Nation*. London: T. N. for W. Lee, 1651.

P[arker], M[artin]. *The woman to the Plovv and the man to the hen-roost; or, A fine way to cure a cot-quean*. London: F. Grove, 1629.

Paget, John. *Mediations of Death Wherein A Christian is taught How to Remember And Prepare for his latter end*. Dort: Henry Ash, 1639.

A Pithie and Short Treatise by Way of Dialogue whereby a godly Christian is directed how to make his last Will and testament. London: William Iones, 1612.

Pliny. *The historie of the world: Commonly called, the naturall historie of C. Plinius Secundus*. Translated by Philemon Holland. London: Adam Islip, 1601.

Plutarch. *The Philosophie, commonlie called, The Morals*. London: Arnold Hatfield, 1603.

Poole, William. *The Countrey Farrier*. London: Tho. Forcet, 1650.

Puttenham, George. *The Arte of English Poesie*. 1589. In *English Renaissance Literary Criticism*, edited by Brian Vickers. Clarendon: Oxford, 1999.

Rastell, John. *An exposition of certaine difficult and obscure words, and termes of the lawes of this realme, newly set foorth and augmented, both in french and English, for the helpe of such younge students as are desirous to attaine knowledge of ye same*. London: Richardi Tottelli, 1579.

Rawlinson, John. *Mercy to a Beast*. Oxford: Joseph Barnes, 1612.

Remnant, Richard. *A Discourse or Historie of Bees*. London: Thomas Slater, 1637.

Shakespeare, William. *The Complete Works of William Shakespeare*. Edited by Stanley Wells and Gary Taylor. Oxford: Clarendon, 1988.

1607. A true report of certaine wonderfull ouerflowings of Waters. London: Edward White, 1607.

Southerne, Edmund. *A Treatise Concerning the right vse and ordering of Bees*. London: Thomas Orwin, 1593.

Stafford, Anthony. *Meditations, and resolutions, moral, divine, politicall*. London: H. L., 1612.

Stow, John. *A Svrvay of London*. London: Iohn Wolfe, 1598.

——. *A suruay of London*. London: Iohn Windet, 1603.

Strype, John. *A survey of the cities of London and Westminster*. London: A Churchill, 1720.

Swinburne, Henry. *A Briefe Treatise of Testaments and Wills.* London: Iohn Windet, 1590.

Taylor, John. *Bull, beare, and horse, cut, curtaile, and longtaile.* London: M. Parsons, 1638.

———. *The Carriers Cosmographie.* London: A. G., 1637.

———. *Iack a Lent his beginning and entertainment.* London: I. T., 1620.

Topsell, Edward. *The Fowles of Heauen or History of Birdes.* Edited by Thomas P. Harrison and F. F. David Hoeniger. Ca. 1613; repr., Austin: University of Texas, 1972.

———. *Historie of Foure-Footed Beastes.* London: William Jaggard, 1607.

Tusser, Thomas. *Five Hundred points of good Husbandry.* London: I. O., 1638.

Vesey, Henry. *The scope of the scripture.* London: W. Iones, 1621.

West, William. *Symbolaeographia. Which may be termed The Art, Description, or Image of Instrucments, Coueneants, Contracts, &c. Or The Notarie or Scriuener.* London: Richard Tothill, 1590.

Whitney, Geffrey. *A Choice of Emblemes.* Introduction by John Manning. 1586; repr., Aldershot: Scolar Press, 1989.

The Young-Man & Maidens Fore-cast; Shewing How They Reckon'd their Chickens before they were Hatcht. London: P. Brooksby, ca. 1680.

INDEX